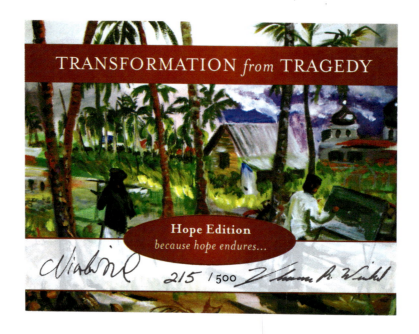

TRANSFORMATION *from* TRAGEDY

Hope Edition
because hope endures…

215 / 500

TRANSFORMATION
from TRAGEDY

Stories of Hope, Faith & Community

After the Tsunami

Thomas R. Winkel & Nicola M. Winkel

with a foreword by
PRESIDENT
GEORGE H.W. BUSH

Bill & Gen,
Many Blessings!

Thomas R. Winkel

4-5-08

Words With a Purpose Press
An Imprint of WordPoint Publishing Group, LLC
Phoenix • Arizona

WordPoint Publishing Group, LLC
Phoenix, Arizona
www.WordPointPublishing.com

Transformation from Tragedy: Stories of Hope, Faith & Community After the Tsunami
© 2008 by Thomas R. Winkel & Nicola M. Winkel
First printing: January 2008

Design and layout by Star Dot Star, LLC.

ISBN-13: 978-0-9794016-0-2

Library of Congress Control Number: 2007933236

This book is printed on acid-free paper.

Printed in China.
12 11 10 09 08 1 2 3 4 5

listen now TO THE GENTLE WHISPERS OF HOPE.

— *Charles D. Brodhead*

foreword BY PRESIDENT
GEORGE H.W. BUSH

In February 2005, I traveled with former President Bill Clinton to the Southeast Asian countries devastated by the earthquake and tsunami. In the spirit of partnership, we came together to draw attention to the cause and to raise funds for the relief effort.

This book, *Transformation from Tragedy*, achieves those very same goals. Thomas & Nicola Winkel are ordinary people who answered the call to serve in extraordinary circumstances. Their experience providing trauma counseling to tsunami survivors and relief workers gives them a unique perspective on the tragedy and resulting transformation in the community of Aceh, Indonesia.

In the pages that follow, you will walk alongside them as they travel into the heart of disaster. Through stories and images, woven together into an unforgettable narrative, you will meet the resilient survivors and see the remarkable efforts of the relief workers. You will come away with a renewed appreciation for the strength of the human spirit.

Be inspired by the role you played in the relief effort as a citizen of the United States. President Clinton and I asked you to support the relief effort, and you gave generously. (Some estimates put private donations from the United States at over $2 billion.) Our military was called on to assist, and they served with distinction. The American people responded with compassion and generosity from their hearts and because of this, people who once thought ill of us now call us friends. When we traveled to the region, we heard story after story of what our fellow Americans had done and were doing. We were never prouder of our great and generous country.

We live in a world and a time when we need to believe that hope endures. Through this story, you will believe.

GEORGE H.W. BUSH
41st President of the United States

YEMEN

BANGLADESH
INDIA

BURMA
THAILAND

SOMALIA

SRI LANKA
MALDIVES   MALAYSIA

KENYA

TANZANIA   INDONESIA

SEYCHELLES

MADAGASCAR

SOUTH
AFRICA

BANDA ACEH

MEULABOH   MEDAN

LAKE TOBA

PHOENIX

SUMATRA,
INDONESIA

The story locations in *Transformation from Tragedy* are coded by color and symbol on the map and throughout the book.

STORY LOCATIONS:

 City of **Banda Aceh**, Indonesia (ban-dah ah-chay)

 Lake Toba, Indonesia (toe-bah)

 City of **Medan**, Indonesia (may-dawn)

 City of **Meulaboh**, Indonesia (muh-law-bow)

 City of **Phoenix**, Arizona, U.S.A.

NOTE: The Aceh (ah-chay) province at the northernmost end of the island of Sumatra was the area primarily affected by the tsunami in Indonesia. Banda Aceh is the capital city of the province.

Countries physically impacted by the tsunami are identified on the map in white.

CONTENTS

what's inside

VOICES FROM ACEH *pages 10 – 31*

Combining illustrations and photographs, *Voices from Aceh* is a unique visual recounting of the earthquake and tsunami. All of the included quotations are directly from survivors we met. The focus is on setting the stage for the stories that follow, while respecting the experience of the survivors.

THE STORIES *pages 32 – 147*

The twelve stories that make up the bulk of the book are written from Thomas' perspective. Picture yourself alongside him…on the airplane flying toward disaster, on the ground meeting a little boy with a changed heart, and on the shores of a picturesque lake helping survivors heal. *Sidebars* throughout the stories feature the perspectives of fellow relief workers, insights on Indonesian culture, the great work done by relief organizations, and other relevant tidbits of information. Each story closes with lessons learned.

PROFILES OF SERVICE *featured throughout the book*

Inserted between the stories are profiles of service titled *Into the Heart of Disaster*. On these pages you will meet a vast array of people who served during the aftermath of the tsunami. These people put their hearts into bringing their own unique abilities to aid the survivors, and each has a story to tell.

A NOTE FROM THE AUTHORS: *Our experience and the story we have to tell takes place in Aceh, Indonesia; however we would like to acknowledge the many other countries affected by the tsunami. Beyond those countries physically impacted (indicated in white on the map on page 6), dozens of other countries suffered loss of life when their citizens were caught in the disaster. Our support and respect go out to survivors and relief workers in all of the affected countries.*

Dear Reader,

We are writing to you from our home in Phoenix, Arizona. It is a warm August morning in the desert, and as we prepare to send this book to press, we want to take a moment to share with you why we embarked on this writing project.

In 2005, we traveled to Aceh (ah-chay), Indonesia, to provide trauma counseling to the survivors and relief workers of the tsunami. We came away truly inspired by what we witnessed and we knew there was an important story to tell. This book is that story—a story of transformation…from tragedy to joy…from trauma to healing…from division to connection. It is an up-close and personal account of lives changed by both disaster and compassion.

Looking back, we can see that many of our life experiences have prepared us for the specific task of writing and publishing this book. We know we were meant to write this story. We also believe you are meant to read it.

We have kept you in the forefront of our minds since writing the very first word. We believed you would be someone moved by human suffering. Someone who seeks uplifting stories of hope to put the world in context. Someone who may have given to the relief effort and had a lingering question of what happened next?

We knew that if you picked up this book we wanted to take you on a journey to the beautiful land of Indonesia; to share with you what it was like to walk amongst the devastation; to bring you face-to-face with the remarkable resilience of the people. Our sincerest hope is that these stories will touch your life as they have touched ours.

In the pages that follow, we invite you to witness tragedy transformed…

Blessings,
Thomas & Nicola Winkel

P.S. Through buying this book, you are directly supporting the tsunami relief effort, as a minimum of 20% of book sales go directly to organizations working in Aceh, Indonesia. (Visit www.WordPointPublishing.com for a list of beneficiary organizations.) We thank you for reading and for your support.

voices from aceh

DEDICATED TO THE SURVIVORS
AND THOSE WHO WERE LOST

ON DECEMBER 25, 2004, FAMILIES AROUND
THE WORLD CELEBRATED CHRISTMAS NIGHT.

Silent night, Holy night

All is calm, all is bright

'Round yon Virgin Mother and Child

Holy Infant so tender and mild

Sleep in Heavenly peace

Sleep in Heavenly peace

ON THE MORNING OF SUNDAY, DECEMBER 26, 2004, PEOPLE THROUGHOUT THE COUNTRIES BORDERING THE INDIAN OCEAN WENT ABOUT THEIR DAY LIKE ANY OTHER.

"It was a normal Sunday."

"The day was beautiful."

"I was headed to the market for vegetables and fish for the evening meal."

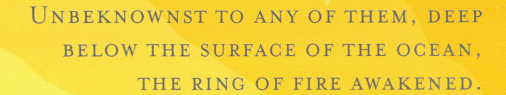

UNBEKNOWNST TO ANY OF THEM, DEEP
BELOW THE SURFACE OF THE OCEAN,
THE RING OF FIRE AWAKENED.

"*The day was unremarkable.*"

"*Many people gathered on the
soccer field watching a match.*"

"*Sunday is my only day off
from work...I was going
to spend it with my family.*"

6.2 MILES BENEATH SEA LEVEL, 750 MILES
OF FAULTLINE RUPTURED, CREATING A
SHOCKWAVE THAT CIRCLED THE
EARTH SEVEN TIMES.

"After the earthquake everything was quiet."

"I was so relieved when the earth stopped moving."

"The earthquake terrified us."

"I had to hold on to a tree because the earth was shaking so badly."

"The earth moved, like taking a sheet to flip the dust off."

"The earthquake knocked me flat."

7:58:53 a.m....earthquake hit off the coast of Sumatra, Indonesia...latitude 3.295 N / longitude 95.892 E...9.3 on the Richter scale

THE QUAKING OF THE EARTH RELEASED ENOUGH ENERGY TO POWER THE UNITED STATES FOR ELEVEN DAYS, SENDING TSUNAMI WAVES AROUND THE WORLD.

"The sound was deafening."

"I thought it was the end of the world."

"We went to the second story of our house. We thought we would be safe, but the water tore through our home."

"We were at the mercy of the water because no one in my family can swim."

"I thought the water would never stop coming."

"Before December 26, I had a wife and children. That day, the water took my family away."

"I have nothing."

Waves as high as 100 feet struck the northern coast of Sumatra, Indonesia...waves and tidal activity reported in Sri Lanka...India...Thailand...South Africa...Canada...Mexico...Chile...Antarctica

Photo by Rodney Rascona

On **D**ECEMBER 26, 2004, THE DAY OF THE DISASTER, THE INITIAL ESTIMATED DEATH TOLL THROUGHOUT SOUTHEAST ASIA WAS 3,000.

THE WESTERN COAST OF THE ACEH PROVINCE LIES LESS THAN 100 MILES FROM THE EPICENTER OF THE 9.3 EARTHQUAKE.

Photos courtesy of GeoEye/CRISP-Singapore

Satellite photos taken before and after the disaster uniquely illustrate the extent of the devastation in coastal areas of the Aceh province.

"*We got in our car during the earthquake because we thought we would be safe. After the earthquake we sat down on the ground. Then we heard screaming...'water, water.'*"

" The water was like a wall. At the bottom of the wall, it chewed at the ground and ate everything. At the top, it was poised like a snake's head, striking at everything. "

No video footage survived (if any was shot at all) of the true scope of the waves in Aceh. Evidence, damage, and survivor accounts suggest the waves reached heights of 70–100 feet in some areas.

Photos courtesy of GeoEye/CRISP-Singapore

Photo by Rodney Rascona

SIX DAYS AFTER
THE TSUNAMI, ON
JANUARY 1, 2005, THE
ESTIMATED DEATH TOLL
ROSE TO 140,000.

IN SOME AREAS, THE VIEW FROM THE GROUND WAS OF EARTH SCRAPED CLEAN. IN OTHER AREAS, THE SCENE WAS AN EERIE LANDSCAPE OF MAN'S CREATIONS DISPLACED BY NATURE.

" The water swept everything away. "

The tsunami waves in Aceh were not only unfathomably high, but also far reaching. When the water receded, a landscape of misplaced objects as small as shoes and as large as cargo ships remained.

Boats in particular created a jarring picture throughout Banda Aceh. Dozens of small boats bunched one on top of another, and huge barges landed in the midst of residential neighborhoods. This fishing boat was left on a city street more than a mile inland.

IN THE DAYS AND WEEKS AFTER THE DISASTER, SURVIVORS SEARCHED FOR WHAT REMAINED OF THEIR HOMES AND THEIR FAMILIES.

U.S. Navy photo by Photographer's Mate Airman Jordon R. Beesley

An Acehnese woman makes her way through the rubble. Survivors like her were left with the task of salvaging what little remained or, in too many cases, going on with nothing but their own survival.

" *The wave took my husband and children.* "

In the aftermath of the disaster, mass graves became a necessity. Thousands of families were plagued with the unending grief of not knowing the true fate of their loved ones, and many clung to the hope that they would be found alive. Along this hospital wall, people posted signs marked *dicari* (di-char-ee), which means *looking for* in Indonesian.

On **January 19, 2005**, twenty-four days after the disaster, Indonesia raised its death toll by 50,000, bringing the total lost to **225,000**.

THE TSUNAMI CREATED AN UNPRECEDENTED
NEED FOR AID; THE PEOPLE OF ISOLATED ACEH
WOULD NOW HAVE TO RELY ON THOSE THEY
PREVIOUSLY DISTRUSTED FOR HELP.

*" I cannot believe that people who live
so far away care so much about us. "*

Before December 26, 2004, Westerners in
particular were not permitted to travel in
Aceh for safety reasons. After the tsunami,
funds, supplies, and personnel poured in from
all over the globe. This worldwide relief effort
caused many Acehnese to reconsider their
negative view of the outside world. Just a few
short weeks after the tsunami, an American
was a welcomed sight to many.

While in Aceh, Thomas had the opportunity to meet
people from all over the world. Here he stands with
Spanish soldiers who were assisting in the relief effort.

AFTER THE WAVES SUBSIDED, THOSE WHO SURVIVED WERE LEFT WITH THEIR SHOCK, GRIEF, AND WONDER...

how will we get through this?

"*I don't see any reason to keep living.*"

It is difficult to comprehend just how many people perished in the earthquake and tsunami; however, seeing the devastation first-hand, it is equally difficult to imagine how anyone survived.

Thomas met this man in a refugee camp. They stand on what used to be his home. When the waves came, the water swept him from his home to the trees in the distance, where he clung to a branch until the water subsided. He lost his wife and all but one child, as well as more than thirty members of his extended family.

Photo by Rodney Rascona

THE TRUE DEATH
TOLL FOR INDONESIA
WILL NEVER BE KNOWN.

the call

Wednesday, February 2, 2005

Today I was thinking that in a matter of days, our lives can **change** *forever. For too many on December 26, that change meant loss, trauma, and pain. For me, the change is displacing the uniformity of everyday life, giving me the opportunity to serve and* **walk with those in need.**

Date:
December 26, 2004

Time:
Morning

Place:
Phoenix, Arizona

Setting:
Thomas is called from the comfort of his living room to the center of disaster in a few short weeks.

 ## departure

The pleasant voice of the airport announcer chimed throughout the international terminal at LAX, leaving messages not meant for me echoing in my ears. It was February 2, 2005, and my journey was just beginning. I felt the effort of preparing physically and emotionally for the journey ahead through to my bones, mixed with a heightened anticipation in my soul of what was to come.

As the boarding call for my Singapore Airlines flight came over the intercom, I gathered my belongings and boarded the plane. Walking down the aisle, I looked for my seat in row thirty-seven. All around me passengers settled in, preparing for the long flight to come, while the flight attendants readied the cabin for departure.

A strange combination of excitement and weariness lay heavy on me. Images of what I knew I'd find on the other end of my twenty-six hours of travel flickered through my thoughts. Although I was exhausted, my mind was already preparing me for my encounter with disaster. Sighing deeply, I reviewed with amazement all that had transpired in just a few short weeks.

"I'm headed back home to China for the Chinese New Year. How about yourself?" my seatmate for the next eighteen hours inquired.

I paused briefly. This was the first time I had been asked where I was going now that my journey had become a reality. I wasn't sure how to respond.

Deciding to go with simplicity, I responded, "I'm headed to Indonesia."

"Ah…business?" he asked.

"Actually, I'm going to help out with the tsunami relief effort."

Now the man took a good look at me, sized me up, and with a very slight tone of incredulity asked, "How did that come about?"

"Well," I said, "quite honestly, it almost didn't."

waking up

Five weeks earlier, Christmas 2004 came and went. I slept well that night, a result of generous portions of "L-Turkeyphan" and too many starches. After rousing from slumber on the morning of December 26, I poured myself a cup of coffee, grabbed a bowl of cereal, and clicked on the television to my customary cable news channel. Instantly, images flashed across the screen that told of the horror earth and water could inflict on humankind. I remained transfixed for hours.

Tremors from earthquakes and tidal fluctuations from tsunami waves are often recorded circling the Earth. In retrospect, I wonder if it was those circling tremors that shook something loose inside of me that day, starting a movement in my heart and mind that would lead me to action.

I watched as the story of the disaster unfolded. What started out as a tragic event with an estimated death toll of 3,000 people quickly spiraled into an epic disaster with victims in the hundreds of thousands. As the death toll rose, so did the feeling that something more than remote compassion was required of me in this situation.

I closely monitored the news for the next several days as video images passed from the hands of survivors who recorded them to those broadcasting them to the world. I listened as survivors told remarkable stories. Day by day, it became evident how fortunate they were to survive at all. I felt the call to action growing in my heart, although I had only a vague idea of what I would do if I went.

I asked myself: Is this call just a need to feel useful in the face of such tragedy?

I sifted through my thoughts and emotions to sort out the source. I prayed simply, *Lord, your will, not mine,*

be done, and waited patiently for clarity.

As the days passed, my patience paid off. I felt a calm assurance and an urging to start walking down this path, even though I did not know where it led.

On New Year's Day, I shared the burden on my heart with my wife, Nicola. She surprised me with her wholehearted support for the idea of going to help. Given the chaos of disaster, any hesitation on her part would have been justified. Yet, there was only support, encouragement, and most importantly, her planning skills. My wife, the planner, stated that I could not just step off an airplane into a disaster area with no specific contacts or purpose (my idea).

Nicola set about researching every possible avenue for me to serve in Thailand or Sri Lanka. She contacted organizations to see if they could use

NICOLA NOTE

My initial reaction to Thomas' desire to go and help surprised me as much as it surprised him. From the first time he brought the idea up, I felt an incredible sense of peace about it. I realized immediately what an awesome opportunity it was not only for Thomas to serve, but also for our community to contribute so tangibly to the relief effort half a world away.

"I FELT THE CALL TO ACTION GROWING IN MY HEART..."

Indo Insights

About Indonesia

Geography: Indonesia is made up of 17,508 islands. Because of its proximity to the equator, the climate is tropical. Indonesia also sits directly on what is known as the Ring of Fire, a zone of frequent seismic and volcanic activity encircling the Pacific Ocean.

Population: There are 238,000,000 people living in Indonesia, making it the fourth most populous nation in the world.

History: Indonesia was a Dutch colony for three hundred years, until declaring independence in 1945.

Language: There are 365 different languages spoken throughout Indonesia. The unifying language is Bahasa Indonesia, which almost everyone speaks.

About Aceh

The Aceh (ah-chay) province is located at the northern most tip of the island of Sumatra in Indonesia. The capital city is Banda Aceh.

Prior to the tsunami, Aceh was fiercely independent, embroiled in a civil war, and closed off from the outside world. A strong Islamic separatist movement wanted Taliban-like laws to govern the land, and the government of Indonesia would not even allow Westerners into the region due to the very high likelihood that they would be kidnapped and killed. The sudden influx of outsiders, both from other parts of Indonesia and from around the world, provided a unique opportunity for Acehnese and foreigners to move past stereotypes and relate to one another on a human level.

me in the relief effort. The answer, in the midst of a disaster of this scale, was that they could not use me. And understandably so. These organizations have trained teams ready to dispatch, and the middle of a disaster is no time for a new volunteer.

We also had to consider our professional life. As a counselor in private practice, it was risky to consider closing my business for a period of several weeks. Would my clients understand? Would they come back to see me after I returned? What would the impact be on my business?

There was also the matter of finances. With no experience in missions, we did not know how feasible it would be to raise support, but we were committed to my being self-supported so I would not be a drain on the relief organizations.

It occurred to us that my flying all the way to the disaster zone at a cost of great time and expense did not make sense unless I specifically needed to be there for some reason. Otherwise, it made far more sense to raise money here and send funds over to support the relief effort.

And how in the world would I communicate with people once I arrived?

So there I was with no access to the relief zone, a caseload full of clients who needed me, and many questions about finances, safety, and logistics. Most of all, I wondered how all this reconciled with the call to action I felt so strongly.

Discouragement crept in even as I tried to stay focused on my purpose.

direction

Then an email arrived. The message welcomed me to come and serve in Indonesia, a country we had previously ruled out. Even though Nicola's family is from Indonesia, the Aceh (ah-chay) province was unsafe for Westerners, so we assumed it was not a viable option for service.

Little did I know that the transcontinental connections of Nicola's parents (who were born in Indonesia and emigrated in the 1960s) would provide me with safe passage right into the center of the disaster zone.

The email message was from an Indonesian man named Peter. While warning me about the chaotic nature of the area and possible rough conditions, he clearly stated that my mental health skills were sorely needed. He also stated that I was welcome in Aceh.

This open door was an answer to my prayers, and the confirmation I needed to move into action. It was January 17, 2005, twenty-two days after the disaster and seventeen days from when I would step onto a plane to Indonesia. In short order, we raised all the funds needed, informed our very understanding clients, and made all of the travel arrangements. My mother-in-law, Vicki, gave me a primer on Indonesian customs and culture to help me navigate.

As I prepared, we watched the news with a growing understanding of what I was heading toward. On January 19, three weeks after the disaster, Indonesia raised its death toll by 50,000 people in one day. In the face of this massive increase, I felt the temptation to let my mind go numb and no longer even try to register the degree of devastation.

But I knew I needed to fight that numbness because we have to *feel* in order to care, and we have to *care* to exercise our humanity.

Five weeks after the disaster, I was on a mission heading toward hell on earth. I had never felt a desire to serve in this way before, and yet, there I was.

I was moved from my comfort zone to relying on the "Great Comforter" in order to bring comfort to strangers that lived a world away.

arrival

The long-haul flight from Los Angeles to Singapore passed uneventfully; at the end of the flight my seatmate and I exchanged email addresses and wished each other well.

The heavy increase in air traffic and foreigners coming through the Medan, Indonesia, airport presented logistical and bureaucratic challenges for those arriving to help. In the first couple of weeks after the disaster, it was not unusual for people to wait days for a flight into the disaster zone.

Messages

An Email of Welcome

January 17, 2005

Dear Brother Thomas,

You are most welcome to visit and come to Indonesia!

My name is Peter, and I am helping out with the relief effort in Aceh and North Sumatra. Your area of expertise is especially in great demand now, so much so that we feel it would not do you justice to confine your efforts to one particular group or hospital only.

You must know, however, that Aceh (tsunami-affected parts) is nowadays in total chaos, so to speak. There are so many teams of volunteers, both domestic and overseas, spread all over that it is becoming increasingly difficult to get good, comfortable accommodations. Even medical doctors must often sleep in tents with a total lack of amenities. It is a virtual war zone!

So, we cannot promise you everything will be easy. Just make sure that you will not become traumatized yourself by what you will encounter in Aceh.

For the time being, we feel you should start with a hospital in Medan, working with the ethnic Chinese refugees who fled from Aceh, having lost all their belongings. Your services are also needed in Meulaboh (on the west coast) and on Nias Island as they are in desperate need of specialists like you.

You are especially needed to train up relief workers who can continue your work of trauma relief after you are gone. As far as translation is concerned, we believe that this should not be so much of a problem.

On behalf of those who support the relief work, may I again bid you welcome.

Peter

P.S. We believe you are heaven sent! And we want to thank you for your willingness to be with us in our time of need!

Each One Counts

A tragedy with the scope of the tsunami can inevitably lead to the question, what difference can any one person make? When provided the opportunity to help someone else, if it is within our means, most of us will take it. But what about helping people who literally live a world away?

The world watched as the terrible drama unfolded on cable news, the mounting death toll casting deep sorrow across the globe, and the survivor accounts making us shake our heads at the mere thought.

After the tsunami, the million-dollar gifts from corporations and celebrities were well-publicized. Somewhat lost in the reporting, though, was the fact that one in four American households made a donation to support the relief effort, raising a total of $2.78 billion. This figure represents a staggering 88% of the total $3.16 billion given by individuals, corporations, and foundations in the United States (*Center on Philanthropy at Indiana University*).

This giving helped make possible the massive influx of supplies, relief workers, and, most importantly, hope to Aceh and all the other affected areas.

Added together, each contribution counts for a world of difference.

A short one-hour flight to Medan (may-dawn), Indonesia, was my last breath of quiet before stepping into chaos. The muggy tropical air enveloped me as I stepped off the plane and walked down the stairs onto the tarmac of the Medan airport.

Medan was the closest, unaffected city to the disaster area. Since the tsunami, the airport traffic had expanded by a hundredfold in a matter of weeks, severely exceeding the capacity of the airport to process the paperwork and adhere to standard regulations for relief organizations. Fortunately for me, the regulations were simplified to something like, "If you look Western, you are here

for relief work."

Unfortunately, stepping into the terminal, I had no clue where to go or what to do. Lines of people stretched before me in a confusing array of nationalities. There appeared to be no direction as to which line I, an American, should stand in. As I watched the scene, trying to decipher the order, my only comfort was the familiar song "Losing My Religion" by REM, blaring from the speakers. Ironically, the song grounded me back into my faith to trust that all would work out as planned.

In the end, I found the correct line, paid my $25, thirty-day visa fee, and off I went.

I braced myself for what might lie ahead. Beyond the passport check and after the baggage claim area, I was faced with a sea of people. Our Indonesian contacts warned me to be very careful about random interactions with unknown ne'er-do-wells who might try to take advantage of a foreigner traveling alone.

Packing for Aceh proved to be a challenge—needing to bring enough to be prepared for unknown conditions while still traveling relatively light. Insect repellent, Kit Kat bars (to stave off the inevitable need for just a taste of home), a tent, water purification tablets (just in case), and a satellite phone were just a few of the items crammed into Thomas' backpack. Fortunately, the tent and tablets went unused, while the satellite phone earned its keep. Here, Thomas stands amid the rubble talking on a satellite phone arranged by HumaniNet™, an organization that provides technology support and assistance to humanitarian organizations.

Scanning the crowd, I felt the stomach lurch that comes from being a stranger in a strange land and the distinct realization that I really did not belong there.

Then I saw it—a small sign, hand written with "THOMAS WINKEL." There could not be a more welcome sight to a weary traveler. Holding the sign was a man ready to greet me and ease my transition into this very foreign country.

Most importantly, seeing my name, I knew I was exactly where I was supposed to be.

LESSONS FROM ACEH
answer the call

When you feel an inkling, see where it leads...

Looking back, it would have been quite easy for me to dismiss the draw to go and help with tsunami relief. I had never done relief work before, and initially there was no clear path for me to follow. Remaining patient and exploring all the possibilities allowed this situation to work. When I talk to people who are considering something new, I always encourage them to follow their inkling and see where it leads.

From the start, don't go it alone...

Gather people around you who can support you, ground you, and encourage you. Whether you're traveling around the world, running a race to raise money for charity, or trying to get through a tough time, having a team of supporters alongside you puts the seemingly impossible within reach.

People are generous when they have something to believe in...

Worldwide giving, as a whole, was astounding after the tsunami, but for me personally, it was humbling. After sharing my call with friends and family, they gave generously to help me get where I needed to go. Even people I had never met gave to support my work in Aceh. It was amazing to see every dollar needed to make the trip happen donated within just a few days.

Plan thoroughly, but also trust...

Prior to departing for Aceh, Nicola and I planned as much as possible. We bought supplies, made contacts, and wrote trauma counseling training materials. In the end, though, we could not possibly have prepared for everything I would encounter. And that's where trust came in. I believed that if I was truly called, I would be made ready for whatever I encountered. And, as you'll see, I was.

INTO THE HEART OF DISASTER:

Profiles of Service

When I stepped off the plane in Indonesia, I had no idea what to expect. My only point of reference was what I had seen on the news, and this story of three young Americans.

On December 26, 2004, friends Ira Lippke, Samuel Lippke, and Scott McAlvany were on the island of Bali, Indonesia, spending their winter break serving at an orphanage in Kuta. Immediately after hearing of the tsunami, they left Bali for Banda Aceh. Their journey took them straight into the heart of disaster.

In Banda Aceh, the main hospital was destroyed and most of the doctors and nurses died in the disaster. Thousands of injured people lined up for help at a makeshift hospital. Though they had no medical training, the need was so immense that the three friends used what basic first aid knowledge they had to help the survivors, even assisting with surgeries.

They dealt with aftershocks, hostility toward outsiders, an environment that assaulted the senses, and endless human need. Ira, a photographer by profession, shares his story in words and pictures.

A Journey of Hope

by Ira Lippke

My brother, Samuel, and I joined my friend, Scott McAlvany, at an orphanage in Bali, Indonesia, for a "humanitarian vacation" during winter break. We had no idea just how life-changing this trip would be.

The day after Christmas, we loaded up all the kids and went to the beach. It's the highlight of the kids' year for them to get out of the city and explore and play at the beach for three days. At the moment the tsunami hit Sumatra, we were literally in the water playing with the kids, but because of Bali's location, the beach where we played wasn't affected at all.

Photo by Samuel Lippke

When we heard about the tsunami, Scott, Samuel, our Indonesian friend Rollie, and I purchased as much medicine as we could get and a water purifier and booked flights to Medan, the closest airport to Banda Aceh.

That night we slept on the floor of the Medan airport waiting to see if we could get a flight to Banda Aceh. In the morning, we hauled all our supplies down the road to the Indonesian Air Force base. After being sent from one person to another, we were able to talk to the governor of Sumatra. He had us handwrite a waiver stating that we were going freely at our own risk and put us on the next plane flying into Banda Aceh. On the plane a soldier asked us, "Okay, why are you really here?" He said, "If it's to exploit the situation for Western gain, you won't come out alive."

Photo by Ira Lippke

Photo by Ira Lippke

When we arrived in Banda Aceh, the stench of disaster and death surrounded us immediately. Huge mass graves were being dug with lines of dead bodies waiting to be buried. The bodies were bloated and burnt—they barely looked human. There was a terrible stench in the air, a mix of broken sewer lines and rotting flesh.

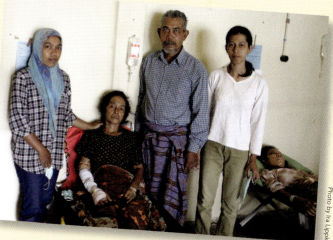

Photo by Ira Lippke

The people of Aceh were so strong and patient. They waited for days to be treated at what remained of the hospital. We were forced to operate without anesthetic. The hospital was filthy and terribly disorganized, yet it was the only option they had. They thanked us over and over for our attempts to help.

We were exhausted and so sick by the time we left, but we felt that with the arrival of the military and medical personnel, we could go home. I don't think we will ever fully comprehend the extent of the destruction and death that we saw. It was one of the most challenging experiences of our lives. We will never be the same.

God's Grace in Abundance

by Scott McAlvany

AN EXCERPT OF SCOTT'S ACCOUNT, AS TOLD IN AN EMAIL SENT TO HIS FAMILY AND FRIENDS.

"I just got back from what was the sum total of the most heartbreaking thing I've ever been involved in…

we saved literally hundreds of lives… which in the big picture is only a small dent in the amount that died while we were there… but every life counts…

first night I performed my first surgery… cut off almost half of a man's arm…it was all dead gangrene flesh…but he is ok…and the doctors said I did exactly what needed to be done…

that first night thirty-six patients died under our care…we worked as hard and as fast…but there was too much to be done and only so many could be reached…

the next day was worse…more died than we saved…kids were screaming as they became orphaned when their last surviving family member would die…

we carried so many bodies…and loaded the trucks to dump them in the mass graves…

third day in…a team of Australian doctors showed up ready to operate…so I got to be the assistant in the operation room all day… we completed nine operations…three were complete amputations…

it was fully a war scene…we were completely covered in blood…people crying…the camera crews had a hard time filming because it smelled so bad…and there was so much trauma in the air…

I cried and threw up and cried more…especially when the little kids lost their parents…

but somehow in it all…God's grace was so abundant…we just charged on…we worked sixteen-hour days with one meal…and still had enough love to give to all…we prayed with every patient, fed them, and changed their clothes…did it all…

Photo by Ira Lippke

four earthquakes hit while we were there...one was a 5.0...

some of the coastal areas are still unreached...and the mortality rates are nearing almost 100 percent in certain areas...the numbers the media have are so underestimated...in any hundred meter circle I could see at least twenty bodies...

end of third day...the Australians showed up in force...and the Marines were scheduled to arrive the next morning...with field hospital gear...

that night...the rebels came down from the mountains and tried to take out the EMT crew plus a few officials...the word on the street was we were next as we were the only Americans in Aceh...so after completing what we felt like was our mission, we made the rounds...fed and prayed and did the final check on all the patients...

we left under armed escort in a military ambulance...they put us on a flight and we were off...

it was by far the single most intense experience ever...we will forever have friends in Aceh..."

Photo by Ira Lippke

Much credit goes to the earliest responders like Samuel, Scott, Rollie, and Ira. Their selfless service and compassion set the tone for the relief effort, assuring survivors, who before considered outsiders enemies, that the world truly cared. The early responders paved the way for the relief workers like me who came after.

– Thomas R. Winkel

Samuel, Scott, Rollie, and Ira at the Indonesian Air Force base in Medan, Indonesia.

Photo courtesy of Ira Lippke

the postcard

Saturday, February 5, 2005

How much is a smile worth? It turns out in the midst of all this chaos and despair, one smile is worth everything. I met a little boy today who smiled, in spite of it all. He smiled because a child who lives 12,000 miles away took thirty minutes to make a special card with a message of caring and hope. And I smiled because I got to be the messenger.

Date:

February 5, 2005

Time:

Late Morning

Place:

Medan, Indonesia

Setting:

Early in his February trip, Thomas visits a refugee camp called "Metal City" with a small team of Indonesian relief workers. He brings survivors tidings from children half a world away.

opening the door

"The people do not want trauma counseling," the doctor stated matter-of-factly. "They want only money to recover from the disaster."

Outwardly, I nodded and did not challenge this statement. Inside, I trusted that I was exactly where I was supposed to be and had faith that I was equipped to help those who suffered. I believed that the possibilities for how I could help the survivors were not yet evident.

Despite the doctor's doubts about there being anything we could do to help, he led our small relief team to a building housing refugees. As we stepped inside and my eyes adjusted from the brightness of the sunny day to the dim interior, I noticed thin foam mattresses covering the concrete floor and marking off areas claimed by the families. These small living spaces seemed a poor substitute for the homes they had lost.

An air of sadness blanketed the room, covering adults and children alike in a melancholy haze. I sensed the reservation they felt toward me, the tall American with the yellow backpack.

One man came forward and shook my hand, likely out of deference for politeness and the social necessity that accompanied a person arriving with the doctor. Other than this gesture, none of the other twenty-five or so adults and children greeted us. The doctor and the rest of the group looked at me to see what I would do.

Instinctively, I knew talking would have little or no effect in this situation. With a burst of inspiration, I crouched down, swung my backpack around, and quickly pulled out a dozen postcards I had brought with me.

"The children where I live made this for you," I said in English, holding out a brightly colored postcard to a little boy. As one of our group members translated, the doctor and everyone else watched for the boy's response. I could only wonder...*would the message be enough?*

postcards across the world

My mind flashed back to the scene a week earlier in an elementary school cafeteria in Arizona.

"Se-la-mat pa-gi," Nicola's mother, Vicki, said to the school children, accentuating the pronunciation of each syllable.

"Se-la-mat pa-gi," the children echoed, a perfectly coordinated replica of my mother-in-law's fluent Indonesian.

"That means *good morning*," Nicola said to the children.

Our mini-lesson on Indonesia, including pointing out the location of the island nation on a map, showing the flag, and sharing facts about the country (300 million people spread out over 17,000 islands), concluded with this brief language lesson.

"You've probably all seen on the news what happened in Asia when the earthquake and tsunami hit," Nicola said.

Tiny heads bobbed in recognition as she continued. "Well, next week Thomas is going to Indonesia to help the survivors of the tsunami and we really need your help. We would like you to make postcards for the children in Indonesia to let them know that people here in America are thinking about them and care."

With a little bit of guidance on subject matter, suggesting they avoid drawing pictures of water, but encouraging pictures relating to Arizona, the children began the art project under the watchful eyes of their teachers.

We walked throughout the cafeteria, observing their sincere efforts. Using prompt sheets showing how to say different phrases in Indonesian, the children wrote messages their recipients would understand. They also drew wonderful pictures and decorated their postcards with stickers of multicultural faces.

Over the course of the morning, most of the classes in the school filtered through the cafeteria for our presentation and activity. In the end, we had over 400 postcards, each one with a destiny and a purpose in the journey ahead.

connection

A week after visiting the school, it was February 5, 2005, less than twenty-four hours since my arrival in Medan, Indonesia. As I packed my belongings into my daypack, the realization hit me: *I've come all this way, now it's time to do what I'm trained to do.*

Before my departure, when people asked how I was going to do trauma counseling, I explained my approach like this: "I'll read as much as possible to refresh my memory on everything relating to trauma and healing. Then, forget everything I know and simply attempt to connect with the people on a human level."

At that moment, facing my first real day of work in the field, and under

Indo Insights

Indo Chat

A sampling of Indonesian phrases the children in Arizona learned to help them make cards for survivors.

ENGLISH	INDONESIAN
COMMON PHRASES	
Good morning!	Selamat pagi!
How are you?	Apa kabar?
I am well.	Saya baik baik saja.
Thank you!	Terima kasih!
MESSAGES	
To our friends in Aceh,	Kepada sahabat-sahabat di Aceh,
Warm greetings from Arizona!	Salam hangat dari Arizona!
The children in Arizona send you a lot of love.	Anak-anak di Arizona mengirim kasih sayang untuk andasemua.
Your friend,	Sahabat Anda,

The students put all their effort into turning their plain, white postcards into bright and hopeful messages of caring, even learning just a sample of a new language to better communicate their messages.

the observation of those who helped bring me here, I sincerely hoped my focus was on target.

I tucked my document folder, which held my passport and other important papers, in the front pocket of my backpack. I put my camera and notebook in the main compartment. As I scanned the belongings in my baggage, I spied a couple of plastic bags filled with postcards. Unsure of whether they would be needed, I figured "why not?" and tucked one of the bags into my pack next to my notebook.

Then off we went, driving through the chaotic city streets of Medan. As we plowed into oncoming traffic to pass a stopped bus, faced off against the opposing traffic at stoplights, and delicately maneuvered the busy streets of motorcycle carts laden with children, I said a quick prayer for a safe arrival.

Pulling up to the refugee camp, I saw first-hand why people dubbed the improvised community "Metal City." Most of the structure's roofs were made of corrugated metal and other available scraps.

We met the doctor in charge of this refugee camp and sat down for our briefing with him. His statement that the people "only wanted money" hung in the air as we continued our conversation. He shared his experiences of attempted "trauma counseling" with this particular group of survivors. I listened carefully to his comments through a translator, gleaning what would help me determine how to best interact with these hurting people.

"These refugee camps were set up in Medan for ethnic Chinese from the disaster zone. There were reports that after the tsunami their Buddhist and Christian beliefs were initially targeted as the reason for God's wrath being

inflicted upon Aceh and they were driven out.

"Many other groups have tried to provide counseling and failed. One group of good, caring people set up sessions, but no one came. Then they tried to incorporate the spiritual beliefs of the survivors by holding the groups at the Buddhist Temple, but again no one came.

"Then they decided to offer 'red packets,' with lucky money inside. Many people showed up for these gifts, but then quickly left before the trauma counseling started."

Now I understood why he came to the conclusion that all the survivors wanted was money. It would have been very easy to be cynical. But it occurred to me that even the act of desiring funds to restart their lives was an attitude of hope for the survivors. They were looking to the future, and this would be key in

their recovery from trauma.

As the doctor spoke, my respect for him grew. He was thrust into a situation that could easily overwhelm even a person very experienced in helping refugees. He handled his charge to care for people, who had literally lost everything, admirably. He was growing wise to the survivors' needs and the understanding of what they were open to accepting from people outside their community and way of life. And yet, I was not discouraged by his belief that there was nothing we could do.

As we wrapped up our conversation and walked into one of the buildings that served as temporary housing for these families, I hoped for clarity for what we could do to help.

It appeared fairly hopeless, given the demeanor of the survivors. Then I had my burst of inspiration. I remembered my last-minute impulse to pack the plastic bag full of the children's postcards.

I held out a card to a young boy. Through an interpreter, I told him about the children where I come from. I told him the children wanted him to know that they cared about him. They hoped he would feel better knowing they were thinking about him.

The young boy in front of me looked at the colorful card with words

Perspectives

Shane Essert

Phoenix resident Shane Essert (pictured above with Phoenix City Councilwoman Peggy Bilsten) was twelve-years-old at the time of the tsunami. He shares the role school children had in helping survivors, and how involvement in the tsunami relief effort changed him.

"Helping with tsunami relief really changed my life and perspective about what it means to have nothing.

When I heard about the tsunami, I was shocked and I felt the need to help. But how, when I'm just a kid? The answer to my question came when Councilwoman Bilsten came to our school to do a presentation about the disaster in Meulaboh, Indonesia. She shared with us how the City of Phoenix was trying to help, and in the end, students from schools throughout Phoenix raised more than $50,000 to help the tsunami survivors.

A few months later Councilwoman Bilsten invited us to a luncheon to meet the Camat (mayor) of Meulaboh. It was so wonderful to see the way God was working through all these people to make this huge relief and rehabilitation project possible. When the Camat got up to tell us what he and his people had lost, I felt so much sorrow for him, I could not explain it.

I love being a part of something like this. It is amazing to see how God works through people to help those in need."

Shane, his mom, Erin, and his sister, Toril, talk with the Camat.

Above photo courtesy of Food for the Hungry

"...CHILDREN IN AMERICA ARE THINKING ABOUT YOU...THEY WANT TO SHARE THEIR LOVE WITH YOU."

Thomas presents a card made by a student from Arizona to a young girl who lost both her parents in the tsunami.

in his native language. Then he smiled at me. He smiled at those in our group. And finally, he smiled at his parents.

The parents were so moved by their son's smile that they brought their daughter over to me and asked if she could have a card as well.

I carefully selected a postcard I thought she would like. It resulted in another wonderful smile.

Then the doctor called over another young girl with an apparent aged and heavyhearted spirit. Not knowing anything about her except what I sensed, I selected a carefully drawn, brightly colored card with a message of friendship and hope.

I got down on one knee and held out the card to her. "A little girl from my home made this for you," I said through a translator. "She and the other children in America are thinking about you, hoping you are well, and they want to share their love with you."

She took the card from me. As she looked at it, her expression remained solemn. Then, a subtle movement. Clutching the card, she drew it to her chest, close to her heart, and smiled tentatively.

The doctor told us he was particularly concerned about her well-being because both of her parents died in the tsunami. She now lived

with her grandfather, but because of his own trauma, it was difficult for him to help her recover from what she had experienced.

Since her arrival, she had been completely closed off to those around her, so much so that her demeanor was no longer that of a little girl. This was the first significant interaction he had seen from her, and even her small smile gave him hope.

Tears filled our eyes as we understood the significance of this small show of emotion from the young girl.

NICOLA NOTE

Thomas shared this story with me over the phone. When he returned and I looked through his photos, I knew right away this was the little girl from the story. The sorrow in her face is heartbreaking. It touched all of us involved in the postcard project that this small gesture of love had a positive effect on her.

I handed out cards to the other children of various ages residing in the building. Each one smiled as they received the simple gift, clutching it as a treasure. Immediately after I gave out the cards, all of the adults, who had been so reserved at first, came swarming over to greet me. There was an onslaught of greetings and thank-you's.

Those cards showed the love and care of a country half a world away better than any words could. The sharing of love from one group of children to another helped bridge vast cultural gaps in an amazing way.

And all of us involved were reminded that "trauma counseling" isn't necessarily all about clinical theories and skills. Sometimes the simple act of carrying and delivering messages of caring and friendship is enough to help those in need.

LESSONS FROM ACEH
make connections

Be flexible...

If I had walked into this camp with an agenda to do trauma counseling in a traditional manner (one-on-one or group sessions), I know it would not have been effective. The doctor was correct. The people did not want that type of help from an outsider at that time. The postcards were a great example of the benefit of being flexible with our approach to helping.

Focus on the needs of those you are there to help...

This seems like a no-brainer, but in the counseling profession and in life, it's something one needs to keep in mind. It becomes easy to get wrapped up in what we think is "best" and not focus on where the person you're trying to help is at and what their needs are. Listening, observing, and discussing needs openly leads to more productive help.

Create a personal connection...

When friends of ours went on a short-term mission to Rwanda, they invited us to sponsor a child from the community they would be visiting. When they returned, they brought photos and video of them meeting "our" child and giving him a book we had sent along with them. Now, even though we've never met him, we have a more personal connection. We hope that the school children who created the postcards had a similar sense of connection to the children affected by the tsunami.

Sometimes it's time, not money...

The postcard project cost very little. It did, however, take the time and effort of many people to pull off. Our friend Cathy, the teacher, coordinated with her fellow teachers to host the event at their school. My mother-in-law, Vicki, and family friends took the time to accompany us to help teach the children Indonesian phrases. Nicola put together a PowerPoint presentation and printed out sample messages. All together, it took a lot of coordination, but in just one day we had gifts for over 400 children. All for the cost of a package of cardstock, some stickers, and time.

INTO THE HEART OF DISASTER:

Profiles of Service

In part, the story of the tsunami is one of people coming together to serve others. One of the most high-profile examples of such a partnership was the joint effort of former Presidents George H.W. Bush and Bill Clinton. The two joined together to spearhead a nationwide fundraising effort.

As part of this effort, they traveled throughout Southeast Asia to see the devastation first-hand and bring further attention to the plight of the survivors.

It just so happened that their visit to Banda Aceh coincided with Thomas' time there, and he recounts the impact he witnessed of the presidents' visit.

Their visit meant the world knew...
It meant the world cared...
It meant the world was acting...
It meant the world had hope for Aceh...
It meant the world.

My Presidents

by Thomas R. Winkel

"The presidents are coming to Aceh. You should go and meet them," Nicola said over the satellite phone, a smile in her voice at the suggestion that it would be so easy.

It did seem unfortunate to miss such an opportunity. I served in the first Gulf War as a Marine, and out of the five key leaders (President George H. W. Bush, Secretary James Baker, Secretary Dick Cheney, General Norman Schwarzkopf, and Chairman of the Joint Chiefs of Staff Colin Powell), I had met Schwarkopf and Baker, but was still hoping to meet the other three at some point. Unfortunately, it wouldn't be quite so easy to garner an introduction, nor did I have the time with my trauma counseling training schedule to seek out a meeting.

I actually knew about the presidents' visit before Nicola mentioned it. A group of Indonesian relief workers had told me just a few minutes earlier. Even in the middle of a devastated area like Banda Aceh, big news traveled fast, and the visit of President Bush Sr. and President Clinton was a very big deal. Not just because of what it was, a visit by two former heads of state to the worst natural disaster the world had seen in recorded history, but because of what it meant.

"Your presidents are coming," an Indonesian relief worker said to me.

"*My* presidents are coming to Aceh?" I clarified.

"Yes, yes, President Bush sent his President Father Bush and President Clinton here to Aceh. They are coming to see the devastation and to help," he said.

Before, during, and after the actual visit, the vibe in Banda Aceh was electric. Not in a manner of visiting stardom, but in a manner of purpose and importance. Their visit sent a message to all those affected by the earthquake and tsunami...*you are important enough to commission two of America's finest to the task.*

To the relief workers, it gave them a sense that they were not alone in their daily struggle to maintain the energy to keep serving. To the survivors, it was a clear sign of hope that the world understood their plight and cared.

To me, it was an affirmation that I was part of something much bigger than my own experience, even if I didn't get to meet the presidents.

First Lady Laura Bush approaches a candle-lit memorial honoring the victims of the tsunami at the Embassy of Indonesia during a visit with President George W. Bush and former Presidents Bush and Clinton in Washington, D.C., on January 3, 2005.

White House photo by Tina Hager

"...I have asked two of America's most distinguished private citizens to head a nationwide charitable fundraising effort. Both men, both presidents, know the great decency of our people. They bring tremendous leadership experience to this role, and they bring good hearts. I am grateful to the former presidents, Clinton and Bush, for taking on this important responsibility and for serving our country once again...

From our own experiences, we know that nothing can take away the grief of those affected by tragedy. We also know that Americans have a history of rising to meet great humanitarian challenges and of providing hope to suffering peoples. As men and women across the devastated region begin to rebuild, we offer our sustained compassion and our generosity, and our assurance that America will be there to help."

President George W. Bush
January 3, 2005

Photo by Gregg Edgar

Former Presidents Clinton and Bush are interviewed on the ground in Banda Aceh, Indonesia. A primary goal of their trip was to draw media attention to the needs of the survivors and the relief effort.

Photo by Gregg Edgar

Former President Bush greets Acehnese children during his visit to Banda Aceh. The visit was coordinated on the ground by USAID (United States Agency for International Development). USAID is the principal U.S. agency that extends assistance to countries recovering from disaster. Since the tsunami, the agency has provided over $400 million in assistance for the relief effort in Indonesia.

Former Presidents Clinton and Bush wave to a large crowd of Sailors and Marines after arriving by helicopter aboard the USS Fort McHenry, while off the Indonesian coast.

U.S. Navy photo

the yes

Saturday, February 5, 2005

The human body's capacity to hold in pain is remarkable.

As I look at this situation, I wonder how these people open their eyes in the morning, how they get up, how they eat, how they speak. The magnitude of the tragedy seems like it should crush them. The pain is irreversible. And yet, it does not break them. They go on.

Date:
February 5, 2005

Time:
Afternoon

Place:
Medan, Indonesia

Setting:
Thomas and a group of relief workers visit survivors in a hospital operated by the Methodist Church in Medan.

survivor

Lucky. The irony of this label hit me as I walked through the hospital ward. Before me lay dozens of patients, all with amputated limbs resulting from injuries sustained during the tsunami disaster. Despite this, they were among the lucky ones to have survived at all.

As our group of relief workers visited the various rooms, our guide explained how these survivors were brought to Medan needing more care than was available in the disaster zones.

Entering one room, the guide led us to a woman's bedside. The intense loss she suffered showed in her eyes. She acknowledged us only briefly, immediately returning back into her private thoughts. She lay in the corner of the five-person hospital room, her head elevated by the hospital bed. The off-white sheets swallowed her slight frame.

She was not covered by a blanket, so the white bandages surrounding what was left of her leg were clearly visible. A multitude of other scratches and newly-healing scars covered her body. Her sunken face had the yellowing marks of deep and old bruises. Unkempt hair hung around her face, a result of the very high humidity. She looked bedraggled and broken, not only in body, but in mind and spirit as well.

A Methodist Church pastor introduced us to her and told us parts of her dramatic story. When she saw our concern, she agreed to tell us her full story of survival and loss.

With no warning system in place and historical knowledge of tsunamis not passed down generationally, people living in Aceh did not anticipate the tsunami after the earthquake. This left most unaware of the approaching danger. Many people sought out vehicles as protection from the earthquake, but what was protective with the shaking earth proved deadly when the water came.

sunday

The morning started like any other Sunday for most Indonesians, with a hustle to the market or a relaxed walk to watch a soccer match. Children played and sisters argued, while fathers ignored the clamor. The day was unremarkable; each person assuming their normal daily tasks.

The violently shaking ground shattered this routine, bringing anyone standing to their hands and knees. In retrospect, the motion was best described as rolling, as if the ground had turned to liquid and you were trying to stand on the surface. Many fell while trying to get their feet under them. Others just stayed on the ground. Most prayed.

Some buildings collapsed, with floors stacking one on top of the other like cards. Others partially collapsed with one end of the building looking perfectly normal and the other twisted and broken, creating a peculiar and unnatural scene. Exposed metal rods, pipes, and wiring, normally unseen, added to the peculiarity.

By all accounts, the earthquake lasted minutes, rumbling across the land with a deep and terrible sound. Some reported screams, others silence, during this time of upheaval. Then it faded and left as it had come.

This now-hospitalized woman had hurriedly gathered her husband and children into the family car. With their home damaged and the looming danger of debris falling, it seemed the safest place to go. Cars are rare in Aceh and she felt blessed to have someplace to protect her family during any aftershocks. She and her husband and their three children sat in the car hoping the metal frame would shelter them from further danger.

The waves came soon after. First, they saw people running and screaming. Then they heard another horrendous sound like the earthquake, but this time they did not feel the ground rolling. Suddenly, the car lurched forward as if hit from behind by a huge truck. But the momentum did not lessen after the initial impact. Their fear of the earthquake was lost with this new terror of the ocean turned black and sweeping over the land. Sure that God's wrath was spilling over them and that they were witnessing the end of days, they felt death very near.

THE EFFECTS OF TRAUMA

A severely traumatic event like the tsunami impacts people in very individual ways. Yet there is some universality in how human beings cope with trauma. Some effects experienced by the survivors included:

CONTINUALLY RE-EXPERIENCING
THE TRAUMATIC EVENT
Many survivors talked about how their mind vividly replayed the events of the earthquake and tsunami over and over like a movie, especially when they tried to sleep.

AVOIDING ASSOCIATION
WITH THE TRAUMA
One common response, especially with children, was fear of any water, even in small amounts like in buckets.

FEELINGS OF HOPELESSNESS
A doctor told me about a husband and wife coming to him and asking for potent medicine. When he asked about their condition, they said, "We have no illness…we want medicine to make us die because without our children, we have no further purpose."

A CONSTANT HEIGHTENED
STATE OF AWARENESS
The energy put toward constant readiness for another disaster often made everyday tasks like cooking and washing clothes seem monumental.

We tried to convey the normalcy of their experiences to survivors. We assured them any person who lived through an event like the tsunami would likely have similar feelings. This seemed to ease their fears of being "crazy" and give them encouragement.

"I CANNOT CRY ANYMORE...MY TEARS HAVE ALL DRIED OUT."

Debris held the doors shut as the wave overtook their car. Crumpling under the weight, the metal shell they had looked to for protection now betrayed them. The mother realized they were trapped deep under the unforgiving black water. Seconds later, the windows burst, the force taking her youngest child.

Her eldest son escaped the car through one of the windows. Only her husband and middle child remained with her in their metal "coffin." Unbeknownst to her, her son made it to the surface to gasp a precious breath and then somehow found his way back down to his family in the darkness. Grabbing his mother, he pulled her through a window, leaving her last child and husband behind.

The shaking of the earth and the force of the water worked in concert to ensure man's works would not last. Deep underneath her, the water stripped the city, her home, down to the earth. With her husband and two of her children gone, this woman clung to her only son, while he grasped at any debris he could hold onto to remain above the water's surface.

They were swept over the houses of their neighbors. The water brimmed with corrugated metal, wood, trees, dead animals, plastic, brick, iron

fencing, concrete, and all the other signs of humanity and civilization. It seemed as if God's intent was to remove the offending presence of mankind and even evidence of their existence. Fear dominated the landscape. Although mother and son lived, they were sure that their deaths were only delayed.

But for them, death would not come. Instead, they were left clinging to a tree as the waters finally receded. They gathered with other survivors at what was left of the mosque. Even as

they believed God had forsaken them, they still wanted to draw near.

The woman and her son waited at the mosque. They waited for a week. They scrounged for food. They drank dirty water. They felt the sun beating down on them. They moved with the shade. They searched out the occasional ocean breezes that brought mild relief from the heat and humidity. They lived surrounded by the lifeless bodies of those they loved, those they knew in passing, and those they would never know.

They watched as other survivors gathered the haphazardly strewn bodies and placed them in rows for identification at the mosque. Later, they watched as bodies were moved a distance away for reasons they cared not to repeat, but that will likely stay with them for a long time.

The mother and son spoke little to each other except of the sequence of events that Sunday morning. Over and over to each other, to others, to themselves, they replayed the scenario, trying with little success to make sense of what occurred and what it all meant.

The woman lost her husband, two children, both of her parents, and all of her nephews. Lower down on the list of importance, she lost her home, car, business, pictures, letters, and all that she had ever known. Within the next few weeks, she would lose her leg after an infection formed at the site of a wound. All these things, unimaginable when put together, created a fortress of pain.

a question

Her story was like other survivors' stories, told in a mournful yet removed manner. As if they were relating a story they had seen in some movie, but with a sadness that only experience could bring. When she concluded, she bent inwards on herself as her face contorted in pain, her arms pressing down as if attempting to force something out of her. We inquired if the pain was from her leg.

"No…the pain is here," she said, patting herself on the chest and stomach. "I cannot cry anymore…my tears have all dried out."

Each of us listening was profoundly affected by her experience. Her declaration hit me like a punch in the chest. I turned my head as tears welled in my eyes. I realized then that my tears, however brief, relieved the physical nature of the emotional experience so that it didn't hurt as much. And I realized that this women before me had no such relief of tears, so her sobbing remained stuck within her,

Indonesia is the most populous Muslim country in the world with approximately 90% of the population practicing Islam. Currently, 97% of the population in Aceh is Muslim. The Aceh region, in particular, is sometimes referred to as the "front porch of Mecca," because historically those making pilgrimages to Mecca would travel through ports in Aceh.

Religion was frequently a part of survivors' perspectives on the disaster. One Acehnese man told Thomas, "God sent the tsunami because we have not been faithful enough." Another survivor, who lost her family, stated simply, "It was God's will."

Many mosques in Aceh survived the waves, as the open architectural design allowed the water to flow through and the sturdy construction withstood the force of the tsunami.

The Torsinas

Our Indonesian friend Tor Torsina and his wife, Ferani, were instrumental in helping to connect Thomas to relief organizations needing assistance with trauma counseling and training.

"Early on, we were involved in sending funds, medicine, and food to Aceh. Then we also became involved in placing relief workers with a desire to help. When Thomas asked me to place him in Aceh, it was not easy, for he was not a physical rescuer, but a counselor of inner wounds. This service was badly needed, but required an interpreter, which we did not have on hand. So I decided to accompany him to Aceh and to help set up training sessions for local volunteers, including the church workers from the Methodist Church in Medan.

We put our love and effort into helping Aceh, and in return we experienced changes within our own perceptions. Prior to departing for Aceh, our friends in Jakarta seriously cautioned me to not discuss spiritual matters, but only be a good Samaritan! But out of deepest sympathy, we dared to offer sincere prayer to the Acehnese, and they willingly accepted it.

The Acehnese were confused as they were taught that foreigners and non-Muslims are infidels, but now they have a different perspective seeing that outsiders are sincere in extending their love. A group of women thanked us for helping Aceh tanpa pamrih, *meaning without condition or expectation of anything in return."*

kicking against her chest and stomach.

How confusing it must have been for her to experience so much and then have her own body betray its natural way of grieving. We had to do something, but what?

"May we pray with you?" my friend asked with great humility.

For a split second I wondered at the appropriateness of such a question and braced myself for the indignant anger. In Aceh, interfaith prayer was not merely discouraged, it was strictly against all social convention and, in most cases, even forbidden.

"Yes," she answered, almost before he completed the question. She said it so simply and quickly that my concerns faded immediately.

She sat up in bed, thrust out her hand and placed it in his just as he was turning his over to offer to take hers.

"Please make prayers for my son, me, and anyone from my family that may still be alive and looking for us."

We prayed for all of these things and, as part of the prayer, also said to her that God is a loving God, that He loves her, and that God is sad with her.

She accepted the prayers and when we opened our eyes, tears of sadness fell on her cheeks, providing quiet relief for the pain in her heart.

We sat with her as the tears came and

in her eyes I saw a glimmer of hope shine forth. Hope and the understanding that, while the process of grieving would continue, she could rest in the knowledge that someday her grief would be less than it was that day.

Then something unexpected happened. The woman looked up at us and smiled. Her face registered a slight air of unfamiliarity with the expression. I suspected it had been some time since she had a reason to smile. With this reclaimed expression on her face, life returned to her countenance as a glow replaced the grayness of her pallor from before.

We continued speaking with the woman and her fellow patients until it was time to go. As we headed out of the room, we stopped at each bed and said goodbye. It was then that a family member of another survivor in the room came over to us.

"Are you Christians?" he asked. In a flash, the reality of being a Christian in the Muslim world of Aceh was upon us. Not knowing why he was asking, we simply answered "yes," unsure of what he would say or do next.

Then, something else unexpected: he also smiled at us.

"I appreciate you coming to visit and that you prayed with the patients," he said.

Nodding in response, we enjoyed a moment of silently acknowledging the massive gulf bridged between our faiths.

Our group left the hospital knowing that prayers were heard. Prayers for tears and healing, prayers to bring love and comfort to those in need, and prayers for peace, both in the heart and in this small corner of the world.

The tsunami was selective in its destruction, much like you might see with the path of a tornado. Here, trees survive with their foliage intact, despite devastation all around them.

LESSONS FROM ACEH
hold on to humanity

Let humility lead...

Walking into a situation like the hospital room, we were not the experts. All the counseling training in the world could not fully prepare us for the degree of trauma and grief the survivors experienced. By connecting out of a place of humility, we were able to acknowledge this gulf, while still offering what we could. In this case, listening and saying a simple, heartfelt prayer were all the "skills" we needed.

We are stronger than we know...

This woman, who smiled in spite of everything she lost, reminded me clearly that the human spirit can endure almost anything, as long as we are cared for and we can find a small amount of hope in a situation.

Make sure you have a buddy...

There is a reason why the buddy system was invented. Many relief workers experienced severe effects from vicarious traumatization when the trauma of the survivors became their own through the helping process. While my time in Aceh changed me, I did not have negative side effects emotionally. I believe this is because in Aceh I had various "buddies" who watched my back while I watched theirs. When I returned, Nicola was my buddy, ensuring that I had an outlet for processing all I experienced.

the boy

Sunday, February 13, 2005

How many times have I sat in front of the television during the news, *watching* *the violence and hatred toward America that exists around the world, and wondered, what would it take to* *change* *the minds* *of those who hate us? What could I say? What could I do? Would kind words and* *goodwill* *be enough? As it turned out, in this case at least, they were enough.*

Date:

February 13, 2005

Time:

Late Afternoon

Place:

Banda Aceh, Indonesia

Setting:

While in the city of Banda Aceh, Thomas accompanies a fellow relief worker to a musical and storytelling performance held to lift the spirits of the adults and children living in refugee camps. While there, he meets a little boy and makes an amazing connection.

Most survivors living in camps in Aceh lost anywhere from 10-90% of their immediate and extended family and almost all of their posessions. Despite this reality, the relief workers put a lot of effort into creating a positive atmosphere. Here, children enjoy a party complete with balloons, streamers, and treats.

photographic memory

"Thomas, take my picture!"

One of my new friends, a relief worker from Jakarta, gestured me over. A young, skinny boy with hollow eyes stood slightly behind her. I recognized "the look" in his eyes well, having seen it on the faces of many of the children affected by the tsunami— an outward showing of the numbness they felt inside.

I smiled at him and the young adult survivor next to him. She may have been his mother, a sister, or even a distant cousin, perhaps the only family he had remaining. His stare did not change. It registered with me that he was hurting even more severely than most of the children.

I gladly set up to take their photo. But as I looked through the viewfinder of the camera, I clearly saw the boy's stare was no longer blank and numb. Now his eyes focused firmly on me. From his expression, I could tell he did not like what he saw. His eyes pierced through the camera at me.

Then, he brought out a beat-up, black plastic toy gun and started pretending to fire it at me. Aggressively shaking his arms and body to accentuate

the firing of the bullets into my body, his young face twisted in hatred. I could only pray for what to do next.

music to my ears

Earlier that day, one of the Indonesian relief workers had told me about a Muslim musician she met on a flight from Medan to Banda Aceh. With the aid of a funny and entertaining storyteller, he was using music to bring laughter back to the Internally Displaced Persons (IDP) camps. She thought we might like to meet and invited me to accompany her to the camp.

A group of us drove through the battered city streets. Only weeks earlier, half of Banda Aceh was torn from itself during the earthquake and tsunami. Most individuals who lived through it believed it was the end of the world. In many ways, they were correct. It was the end of the world as they knew it.

A significant part of this change was the influx of regular people turned relief workers both from all over the world and from other parts of Indonesia. Many of the workers I served alongside were young twentysomethings from a handful of the 17,000 islands that make up the nation of Indonesia. These young people wore Western clothing and, more often than not, listened to Western music. Not the Mel Tillis, Hank Williams kind of Western music, but the kind you'd see on *American Idol*. *Indonesian Idol* was all the rage, and a couple of them dreamed about singing on the show. Despite their humble nature, I encouraged them to sing loudly. The songs they sang were inspirational, and one thing we all definitely needed was more encouragement.

We arrived at the IDP camp where the musician was performing that day. I felt comfortable walking into the camp, but I immediately sensed something was different. My previous visits to other camps were generally peppered with kind greetings and smiles from the residents interested in the foreigner. Indonesians are known for their smiles. They smile to express kindness even in times of great pain or great discomfort. They smile all the time, and I mean all the time. My face would often hurt by the end of each day from so much smiling!

As we walked into this particular camp, I noticed right away that there were no smiles. Hundreds of people, grouped in what remained of their family units, leaned out or hovered outside the various structures they had compiled from palettes, tents, and tarps. Their faces wore both the weariness of disaster and a cloak of

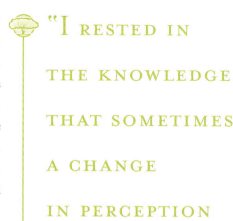

"I RESTED IN THE KNOWLEDGE THAT SOMETIMES A CHANGE IN PERCEPTION TAKES TIME..."

suspicion about me. Quite suddenly, I remembered I was walking through a part of Indonesia that only a few short weeks ago was as anti-American a place as you could find in the world.

Who do they think I am? I wondered. Former American Presidents Bush and Clinton had been to Banda Aceh recently. Perhaps they thought I was some leftover dignitary from their visit? They likely thought I was some Westerner of importance, due to the sizable gathering of Indonesians that accompanied me.

Feeling the pressure to represent America well, no matter who they thought I was, I smiled even more. But it seemed to make no difference. I rested in the knowledge that sometimes a change in perception takes time and hoped their suspicions would ease.

As we walked, distant Indo-Arabian chanting music grew louder until we arrived at the end of the long driveway leading from the congested street to a stately building. Most government buildings in Indonesia are large white structures with vast open areas in the center surrounded by plastered brick and ironwork walls and are guarded by

Survivors gather in the government building compound turned IDP camp for a performance by the musician and storyteller.

armed security personnel.

I found myself thinking of the Taj Mahal, or, more appropriately, the Hagia Sophia in Istanbul, when I first caught a glimpse of an Indonesian governmental compound. They are large impressive structures and their offset nature lends a sense of exclusion to them—as if they were separate from the typically disparate masses and haphazard structures that they govern. Clearly, these were confusing times, however, because the refugee masses and haphazard structures they had built completely filled in the compound's previous remoteness.

The driveway ended at the steps of the building, where a large crowd gathered, enthralled with a performance. We could see musicians playing in the background, while the animated storyteller had the children entranced. His expressive face told the story of a black wave that came and made the characters in the story very sad. The children reacted enthusiastically, booing at the wave and then cheering as the storyteller wove the narrative back to a triumphant conclusion. He seemed a master and held all of us, young and old, spellbound with his performance.

It was as the storyteller concluded and the singing resumed that my friend asked me to take her picture with the young, angry boy.

Looking through the viewfinder and seeing hatred bubbling forth as he pantomimed killing me, I wondered what he had been told about Americans. That we are evil? Depraved? That we all deserve to die? I prayed for divine

guidance on how to reach this child and simply make a positive connection.

Setting the camera down, I said softly, "No, no!" Gesturing to myself, I said, "I'm good, I'm okay." Then I gave the thumbs up sign, while lightly patting myself on the chest. I smiled, willing him to accept my gesture. My friend crouched beside him and, looking at him face-to-face, told him I was from America and I was here to help. His expression did not change with this knowledge.

It became clear to me that this was likely one of the IDP camps not frequented by Western relief workers. Therefore, the inhabitants had not had the exposure that could dispel some of the fundamental negative beliefs about Americans. I realized I had become a dignitary after all!

I prayed to be able to reach this young boy. I could hardly keep the images of 9/11, suicide bombers, and jihadists from running through my head, but I needed to. I focused my attention on this child, knowing that he was just that, a child, who had probably never interacted with a Westerner before.

I knelt down, smiled through my eyes, and held his stare. My friend knew to translate what I said.

"I like your shirt," I said. Though worn with many deep stains, an image of Spiderman still showed. "If I were your age still, I'd like to wear a shirt like that."

He said nothing but kept my gaze. I could see him thinking, perhaps trying to reconcile all he'd been taught with the reality of me, right in front of him, liking his shirt. I gently patted him on the back, calling up all the love and kindness I could through my facial expressions. I hoped it was enough.

The musician started again. The boy and young woman moved to go watch from a closer vantage point. I watched him go. He looked back one time. I could tell he was still thinking.

greetings

I turned my attention to the music, which was amazing. One of my friends aptly described it as a mix of Indonesian, Middle-Eastern, and Reggae sounds. I added that there were strong undertones of Miles Davis/John Coltrane. He laughingly agreed. I enjoyed the opportunity to simply feel the joy of music with my fellow relief workers and, most especially, with the survivors.

After the concert, we ushered ourselves over to meet the musician. I was oddly excited at my first backstage experience, even if the "backstage" was

CHANGED VIEWPOINTS

When people ask, "Did you ever feel unsafe in Aceh?" I tell them about my conversations with survivors.

Many survivors shared their darkest emotions with me…the guilt they felt for not saving their family, the despair of losing a child. Sometimes our conversation would drift to all the changes in Aceh since the tsunami.

I would ask them, "What would have happened if I walked down the street here on December 25?" Without pause, they would answer, "You probably would have been killed." And, they would add, "I would not have disagreed with that action." That these people were telling me their deepest emotions, when just a few weeks ago I would have been an enemy target in their eyes, shows the remarkable ability for perceptions to change because of positive action.

That's not to say that no one in Aceh had reservations about all of the outside help. I also encountered survivors who distrusted the help given by Westerners, and I experienced some tense moments. Overall, though, survivors expressed their amazement that people from so far away cared so much for them.

"How beautiful upon the mountains are the feet of him that bringeth good tidings, that publisheth peace." Isaiah 52:7a

Acehnese Greetings

Handshake - The traditional handshake in Aceh consists of a light handshake, followed by each person gently touching their own heart, which represents the sincerity of the greeting.

Terima Kasih - This is the traditional way of saying thank you in Indonesian. Literally translated, it means *receive love*.

Smiles - In general, the Indonesian people smile often and with enthusiasm. Importantly, they also tend to smile when uncomfortable, so paying attention to their body language is key.

Titles - As a sign of respect, men are referred to as *Pak* (pah) followed by their name (e.g. Pak Thomas), while women are referred to as *Ibu* (ee-boo) followed by their name.

the lobby of a government building. My friend introduced me to the musician. We each acknowledged that there was a lot to be done and all efforts were needed, even if they seemed a mere drop in the bucket. We exchanged appreciation for the work that the other was doing.

After this encounter, my friend mentioned going back outside to meet the storyteller, who was mingling with the crowd. We left the building and found that the crowd from the concert remained intact, with the people now milling about. As we came out, the demeanor of the people transformed from suspicion to smiling faces as they rushed forward to greet me.

Quickly surrounded by the crowd, my friends moved ahead, leaving me to move slowly through the mass of people. Shaking outstretched hands with the traditional Acehnese handshake, I made eye contact while greeting each individual. At some point, I noticed I was shaking each person's hand multiple times. As the number of people multiplied, I finally resigned myself to cutting each person off at three handshakes!

I could see my friends smiling and laughing at my "problem." I could not help but smile as well. I realized how

Photo by Vicki Oei

our brief encounter with the musician had given me a pass with the masses. Their interest and excitement replaced their previous suspicion and stares.

As we started to work our way slowly out of the adult crowd, we came to the throngs of children. Along with the standard forms of greeting, the children were much more apt to grab our hands or forearms. They usually held on for a few seconds and then let go after they had been recognized.

There were so many children, but then I saw eyes I recognized. It was the boy who pretended to shoot me. This time he looked at me straight on and smiled. Taking hold of my hand, he would not let go. As my friends guided us through the children over to the storyteller for a brief introduction, he held on.

Acknowledging the crowd's positive reaction to each of us, and not wanting to divert from those in need, the storyteller and I gave each other a fitting smile and knowing nod. At that point, it was time for us to go. Leaning down to the boy as he held my hand, I smiled again.

"I'm glad you found me," I said.

He smiled up at me innocently and continued holding my hand. I glanced around and noticed my friends were watching. We all knew this was special. Following my friends, the boy and I walked hand-in-hand for quite a while toward the gates. Other children would run up happily, hold my free hand for a while and then fall back while we walked through the compound. I kept looking down at the boy. He would look up at me confidently with those pained eyes and smile. I had tears in my eyes and also shared his smile.

He insisted on walking me all the way to the vehicle, even assisting the driver in closing the door. His eyes held mine as we waved and I said a final goodbye. As the car pulled away, I felt thankful and humbled for such a positive answer to my very simple prayer.

Thomas walks through the IDP camp with his new friend.

LESSONS FROM ACEH
seek clarity

When uncomfortable in a situation, take a moment...

When I first arrived at the camp to the glaring stares of the residents, I had to remind myself to relax. My very human nature told me to leave because clearly I was not welcome there, but my instinct told me it wasn't dangerous, just uncomfortable. Living with the discomfort for a little while gave me the time to immerse myself in the setting, and wait for a window of connection to open. As humans, we seek comfort, but sometimes we have to wade through a little discomfort to get there.

Respond, don't react...

My alternate reactions to the little boy pretending to shoot me could easily have included anger, offense, or fear. Anger at whoever gave him that toy gun. Offense at the hostility when I was only there to help. Fear of whether his actions represented the feelings of people who were capable of bringing me harm. I'm grateful that I was able to set those reactions aside. By pausing, I was able to respond to him, a little boy, instead of reacting to what he represented.

Ask for guidance...

This is a situation where I needed help. My prayer for insight on how to respond to the little boy helped me see the situation more clearly. This was a great lesson in asking for help, whether from those around us or through prayer.

Be an ambassador...

I realized fairly early on that working with survivors in Aceh was a form of diplomacy. With this understanding, I was very conscious of the need to represent my country and the outside world in general in a positive way. My rule became, when in doubt, smile.

Into the Heart of Disaster:
Profiles of Service

One of the most significant aspects of the relief effort was the contribution of the United States military, which provided wide scale relief to the Aceh region in the weeks and months after the disaster.

This effort provided an incredible opportunity to tangibly share the heart of the United States, not just as a country that brought aid, but as individual service people who brought supplies, expertise, and most importantly, compassionate hearts.

Helping Heroes
by Thomas R. Winkel

My dog tags always go with me when I travel. After four years in the United States Marine Corps, they became a familiar comfort, whether in the sands of Saudi Arabia or hiking the canyons of Arizona. When it came time to depart for Aceh, though, I reluctantly left them behind. In the weeks after the disaster, many things were still unsure. Would the distinctive rattling around my neck garner suspicion in a society so distrustful of outsiders? Would it distract those who I was trying to help with questions about my intentions? As an American traveling to a place where only weeks before Americans weren't welcome, I was keenly aware of the need to fly under the radar, at least until I could assess the situation for myself.

Fortunately, sometimes our concerns, no matter how well grounded, dissolve in an instant. Upon arrival in Indonesia, I was told of rumors circulated in Aceh immediately after the disaster that the the U.S. Military detonated an atomic weapon at the fault line to destroy the Acehnese. Thankfully, the person went on to explain that after seeing the U.S. Military's assistance, and more specifically the caring of individual service people, the Acehnese did not believe this anymore; their distrust was replaced with appreciation and admiration for our armed forces.

This remarkable transformation in perception is a direct result of the actions and caring of our military personnel, widely considered "Helping Heroes" by the Acehnese. One man, whose surviving family members' lives hung in the balance after the disaster, told me, "to all of the military people, we will always have them as a part of our family. They could come here and live with us." I told him I would pass his message along.

> **" *The helicopters gave us food when no one else could.* "**

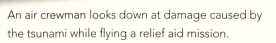

An air crewman looks down at damage caused by the tsunami while flying a relief aid mission.

U.S. Navy photo by Photographer's Mate 3rd Class Jacob J. Kirk

Aviation Warfare Systems Operator 2nd Class Dave Matthews of Freehold, N.J., hands relief supplies from an SH-60F Seahawk helicopter to an Indonesian man.

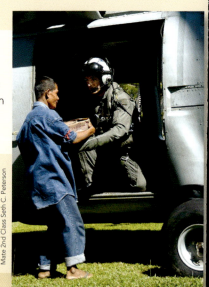

U.S. Navy photo by Photographer's Mate 2nd Class Seth C. Peterson

An MH-60S Knighthawk helicopter idles on a provincial road outside of Banda Aceh as air crewmen pass out relief supplies.

Hospital Corpsman 1st Class Brent Snyder hands a box of bottled water to other members of Navy Environmental Preventive Medicine Unit Six (NEPMU-6) during distribution at an IDP camp. The water storage tanks in the Banda Aceh camp were contaminated, so bottled water was distributed to help the people until new tanks could be delivered.

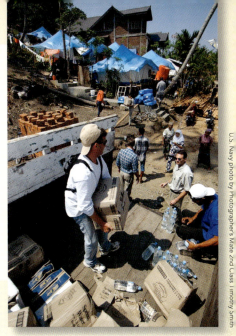

" I love the Marines!"
(Semper Fi!)

Navy engineers assigned to the USS Abraham Lincoln and the Australian Army work together to get a portable generator running at a hospital in Banda Aceh.

Landing Craft Air Cushion (LCAC) vehicles, which are capable of transporting more supplies than helicopters in a single trip, deliver needed materials and supplies to Meulaboh. The LCACs were part of the Bonhomme Richard Expeditionary Strike Group that operated off the coast of Indonesia in support of Operation Unified Assistance, the humanitarian operation effort.

The Military Sealift Command hospital ship USNS Mercy navigates alongside the USS Abraham Lincoln after arriving on station near Banda Aceh.

Lt. Shawn Harris, assigned to the USS Shoup, carries an injured boy from a helicopter to a triage site set up by various relief groups on the Sultan Iskandar Muda Indonesian Air Force Base in Aceh.

" The U.S. Military people cared for us enough to keep us alive. "

Cmdr. Suzanne Clark, a U.S. Navy Nurse, assigned to the hospital ship USNS Mercy, gives a lecture on diabetes to Indonesian nursing students at the University Hospital in Banda Aceh.

Lt. Cmdr. Carma Ericksonhurt, assigned to the hospital ship USNS Mercy, explains her rank insignia to Indonesian military and civilian nurses after instructing the proper applications of CPR at the Tentera Nasional Indonesia Military Hospital.

Hospital Corpsman, Seaman Joshua Scott, assigned to the hospital ship USNS Mercy, entertains Indonesian children with his portable CD player at University Hospital in Banda Aceh.

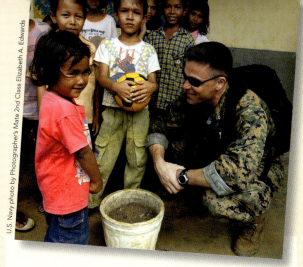

Lt. Col. Andrew Wilcox, one of several U.S. Marines assisting the United Nations, befriends Indonesian children near the town of Glebruk.

"We will never forget those who helped us."

A young Indonesian girl clasps a teddy bear given to her by Sailors assigned to the USS Abraham Lincoln. Sailors collected and donated over 250 of their own toys (many of them mementos from home) to give to Indonesian children.

Indonesian children give a thumbs-up in celebration of relief aid.

A local girl from Banda Aceh proudly plays with her American flag.

U.S. Marines Lance Cpl. Kenneth Gregoire and Lance Cpl. Thomas Reed talk to local Indonesians while at Sultan Iskandar Muda Air Force Base in Banda Aceh.

the durian

Friday, February 11, 2005

I've felt "at home" in the desert of Arizona and the sands of Saudi Arabia. Now, I can add the tropics of Indonesia to that list. As a stranger dropped into the middle of this landscape full of devastation, I didn't expect to find community. Perhaps because I wasn't thinking about my beloved durian!

Date:

February 11, 2005

Time:

Night

Place:

Banda Aceh, Indonesia

Setting:

An unusual fruit serves as an unexpected connection point between Thomas and his fellow relief workers, transforming shared work into friendship.

sending

"What are those?" my friend asked, pointing at the spiny round forms sitting on a table on the back patio. The two good-sized pieces of fruit looked fairly innocuous, despite their thorny exterior.

Glancing out through the sliding glass door at them, I smiled gleefully and arched my brow with anticipation as I answered, "dessert!"

The night before my departure for Aceh, Nicola and I gathered together our family and friends for a traditional Indonesian meal of celebration. I looked forward to spending time with everyone before the long journey ahead. I also looked forward to introducing my fellow American friends to a great love of mine...the durian fruit.

Any Indonesian will tell you that the durian plays a divisive role in human society. This is due to its fragrance, either smelling like rotting garbage or like a sweet rose, depending on your point of olfaction. Upon first scent, people are divided into those who love it and those who can't stand it.

The difference truly is that dramatic, and it can cause everything from domestic disputes, where the wife bans the husband from giving her a good night kiss due to his stinky breath, to being cast from hotel rooms by the management for ignoring the posted signs banning the discriminated fruit for its strong and controversial smell.

After dinner, I announced loudly, "anyone who wants to try durian, come out to the back patio." Some came out because they truly like trying new things. Some came out because they wanted a challenge. Others came out to watch the whole scene unfold.

I rolled the large fruit around in my hands, while the crowd looked on. Turning it toward me, I found a small crack at the base and, with my thumbs, pried open the prickly casing of the fruit. I admit that I quite enjoyed the horror-struck expressions of those who just a scant few minutes ago were very excited to "try something new."

The fragrance filtered out through the crowd. I cut the chunks of durian into small pieces and started passing them around. Smiles and laughter abounded as we bonded over this new experience. Despite only a handful of them falling into the "sweet rose" category, the sight of wrinkled noses

NICOLA NOTE

I definitely fall into the "rotting garbage" camp. Despite this difference, our relationship endures, as Thomas is kind enough to eat his durian on the back patio.

and shaking heads sent me across the sea with warm joy in my heart and a story to share with new friends I would meet.

great care

"Have you had enough rest, Thomas?" the young Indonesian man said to me.

"We want you to retire early so that you have enough energy for the training tomorrow," his equally young counterpart said with great care in her voice.

I thanked them for their concern and said goodnight as they shooed me off to sleep. As I walked toward my room, I smiled at the exchange. Even though I had quite a few years on my new young friends, they made a great effort to ensure that I was taking good care of myself in the days and nights between trauma counseling trainings. Their concern came across in a loving, almost parental manner.

I tried to hold on to the warmth of their concern as I readied for bed. That warmth served as a scant barrier between me and the images I could feel lingering close by. The moment the lights went out, with all other senses dulled and the busyness of the day set aside, my mind was free. And this is when the mind's ability to take a story we hear and reproduce it as a movie in our heads came into play.

As relief workers who arrived in the weeks after the disaster, we saw the dead. We tried to console the inconsolable. We had to communicate the desperate conditions to those outside when all we wanted to do was forget.

Then a curious thing happened. While there, we started living the same lives as the survivors. The images of destruction that the survivors faced became ours. Not just because we lived in the same shattered environment, but also because we internalized their accounts of the horror of the disaster and the grief of the aftermath. Each of their stories and accounts became a part of our personal movie, ready to play the moment the mind sat idle. This replaying is a classic symptom of trauma.

Some of the relief workers who saw the worst were affected only slightly, while some who saw little were affected greatly. No matter the degree, one of the best vaccines for the trauma virus is support and encouragement from those closest to you.

Settling into bed, images of destruction and despair rolled through my mind as I knew they would. Trying to distract myself long enough to fall asleep, I thought of the email I received earlier that day from Nicola. It was a compilation of messages from

National Pride

The response of the world to the tsunami disaster is often touted, and deservedly so. Somewhat in the background, however, is that despite the vastness and diversity of the world's relief effort, the response from Indonesians themselves far outweighed any other nation.

They came from all parts of the island nation to serve in whatever capacity they could, sacrificing their own comfort to help those in need.

As I worked alongside a group of Nationals in Banda Aceh, I found myself impressed and amazed daily at their incredible dedication, and the joyful spirit with which they served. No matter the degree of devastation, death, and destruction, their manner remained upbeat and their purpose resolute.

In the short time I served with them, I witnessed this group of individuals transform into a community through collective experiences and purpose. This mutual support and camaraderie helped them serve others under the most difficult of circumstances.

Messages

Sending Encouragement

Throughout my time in Aceh, friends and family sent me email messages like these to encourage me through the challenging circumstances.

"We think of you often and pray for you as you start each day."

"I feel as if we all are with you. Like we are in your backpack, your canteen, your heart. We are proud to know you, Thomas."

"Traveling mercies, my friend."

"Thomas, I can't imagine what you are experiencing, watch for the miracles…"

"You are touching a lot of lives…more than you will ever know."

"May God watch over you constantly and bring you comfort."

"We're continuing to pray that you will stay healthy in body and spirit."

"God speed you on your mission and give you the strength to carry it out. The world is behind you!"

The support I felt encouraged me greatly, especially knowing that these prayers and bits of encouragement were not just for my well-being, but for the entire relief effort.

"I tell you the truth, whatever you did for one of the least of these brothers of mine, you did for me." Matthew 25:40 (NIV)

---

our close friends and family. They sent messages of encouragement for my soul, of excitement for the work I was doing, and prayers for the challenges.

All at once, I felt the vast distance between my present circumstance and the familiarity of friends and family distinctly. The pull between being in Banda Aceh and helping people, as I knew I was meant to do, and being back home, in my own comfort zone, stretched tight.

Finally drifting off to sleep, I looked forward to feeling at home again soon.

receiving

Before I left for Indonesia, my mother-in-law gave me a primer on Indonesian culture to aid my interactions with those I would meet. "They will ask you two questions: *Are you married?* and *Do you have children?*" She seemed so sure of this, but I wondered if that would really be true across the board. It was a good lesson in trusting your mother-in-law, as more often than not, those were the first two questions my new Indonesian acquaintances asked.

Subsequent questions inevitably revolved around my immediate comfort level with the food. I enjoyed seeing the astonished faces of the national relief workers when they discovered my great affection for their favorite foods. They inquired again and again how I even knew about the hot chili paste *sambal*, durian fruit, and the other Indonesian fare we conversed about.

I told them of my in-laws, who emigrated to North America from Indonesia in the 1960s. After years of exposing me to every dish, sauce, and delicacy they could, their assessment of me was that I may have the build of an American, but I have the tongue and stomach of an Indonesian. My new friends loved this description.

I slowly realized that as I shared these things about myself, my Indonesian friends could see that I cared for and respected them and their culture. I think that they spend so much time watching and thinking about American culture that they can't imagine that Americans would be interested in their life! I found it a little sad that they were so amazed that an American would value their culture. I tried to assure them that many enjoyed the chance to learn about Indonesia, sample the food, and understand the hearts of the people.

"Just the night before I left," I shared, "we had a special dinner with family and friends. At the end of the night, we had two huge durian!" This proclamation was inevitably greeted with great excitement, followed by great curiosity.

"How do you get durian in the States?" one friend asked. "You can't grow it there…can you?"

I told them tales of me, a tall, Caucasian American, walking the aisles of our local Chinese market. As I searched through a huge bin of imported durian to find just the right one, tiny Asian grandmothers looked on with smiling approval at my selection.

I told them about sitting around our patio table with my father-in-law, Nicola's grandfather, and young Spencer. Four generations of males ranging in age from eight- to eighty-six-years-old, all eating durian together.

Then I would then tell them that Spencer is my dog, which would send the whole room into fits of laughter and disbelief.

"Your dog eats durian!!!" they exclaimed.

"He takes after me," I replied, which was followed by more laughter.

Fortunately, in Indonesia, the durian did not create divisiveness between me and my new friends, it created unity.

At one point, someone had a brilliant inspiration. "We have to have a durian party!"

"Absolutely!" I agreed.

As it turns out, it took quite an effort on the part of one Indonesian young man to pull such an occasion together. After repeated forays into the undamaged half of Banda Aceh, he managed to gather up several small durian.

Late one evening, I gathered with a dozen other relief workers on a porch. The darkness of night covered the destruction, allowing us to detach from the trauma we witnessed by day, while the glow of the lights created a feeling of security.

As I stood there, enjoying a special treat with my new friends, any sense of unease faded into the darkness, leaving me with only the scent of durian, the sound of laughter, and an amazing sense of belonging.

A Nasi Kuning, a traditional Indonesian celebration feast, was held for Thomas on the eve of his departure for Aceh. The centerpiece of the dinner is a cone of yellow rice, spiced with turmeric.

Indo Insights

Indo Cuisine

Indonesian food may not be as well known as perhaps Thai food, but it has plenty to satisfy a big appetite.

Sauces - Indonesian sauces, tasty enough to put on almost anything, include a peanut sauce with a kick, a sweet soy sauce, and *bumbu* (boom-boo), a blend of spices.

Gado Gado - So nice they named it twice! A vegetable dish paired with a spicy peanut sauce.

Rendang - A sweet/spicy stewed meat…tasty over a bed of rice.

Sambal - Everything from mango to shrimp is paired with this spicy chili paste (beware…not for the faint of heart or palate!).

Drinks - A variety of unique Indonesian drinks like *Es Cendol* (s chen-doll), a sweet coconut dessert drink, add a special something to any meal. My favorite drink from Aceh? An avocado and chocolate shake…yum!

Selamat makan! (Bon appetit!)

> "...I MAY HAVE
> THE BUILD OF
> AN AMERICAN,
> BUT I HAVE THE
> TONGUE AND
> STOMACH OF
> AN INDONESIAN."

 How to impress your new Indonesian friends (a.k.a. How to eat a durian)

1. Select a ripe fruit

Look for durian in Asian supermarkets. In North America they are often frozen to preserve freshness. Strong debate exists on how to choose a ripe fruit. I look for a durian with small cracks in the hull.

2. Open the durian

Look for a thin crack in the hull and use your thumbs to split it open. (Careful of the spines!)

3. Peel back the hull

Remove each section of the hull to reveal the pieces of creamy, yellow fruit.

4. Enjoy!

Durian is best enjoyed with good friends (and outside if your spouse does not enjoy it!).

Bobby, a national relief worker Thomas met in Aceh, drew this caricature of Thomas and a durian.

A group of Thomas' new friends celebrate during the durian party, a much needed time of levity in the midst of challenging relief work.

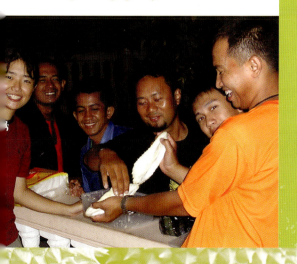

LESSONS FROM ACEH
create comfort

Fish stomach is not comfort food (for me)...

I am an adventurous eater by nature, so not much generally throws me. However, a plate of fish stomach (a specialty in Indonesia) did give me pause during a meal in Medan. Only a few days removed from home, in a foreign country, and working under very challenging circumstances, adventurous eating was not really a high priority for me. Understanding that in most cultures it is disrespectful to not try food served to you, I took a deep breath, smiled enthusiastically, and downed the sauce-covered piece of stomach (which, by the way, had a very spongy consistency). Ironically, when Nicola mentioned my unusual meal to an Indonesian friend later, her reply was, "ooh...I love fish stomach!" showing that comfort is all relative.

Always bring a little taste of home on the road...

You never know what culinary adventures you'll encounter on the road. In fact, I don't think I've ever had a time of traveling overseas when I didn't crave something familiar at some point. (Okay, maybe not in Italy!) The lesson is this: always bring a little something, whether it's trail mix, granola bars, or even a Snickers, to get you by if there comes a traveling day when the menu doesn't seem very appetizing. A dose of comfort food can also stave off an episode of homesickness.

Use your downtime...

Downtime, whether from work, stress, or life in general, is an important factor in maintaining good mental health. When we effectively use downtime, it strengthens us and makes us better equipped to manage the stressful times. It is especially important to really use the time to recuperate. This might mean resisting the urge to worry about issues and perhaps disconnecting from reminders like cell phones, email, and even people, even if just for a short while.

the one

Saturday, February 12, 2005

As relief workers, how do we get through the day when there are so many people who need help*? When the* sorrow *is so great? When it seems like the loss is so profound that nothing we do or say could possibly make it any better? Today I was reminded, we* look for the one*.*

Date:
February 12, 2005

Time:
Midday

Place:
Banda Aceh, Indonesia

Setting:
Thomas conducts a training on trauma, helping to equip relief workers and helping one survivor begin to heal.

 anger

The door to the restaurant swung open and flooded the room with bright sunlight. The noise of a nearby jackhammer, separating the remains of a fallen building, interrupted my sentence. I paused for a moment as a man walked into the room. With his head down, he walked directly to an empty seat without making eye contact.

As I continued my training, a fellow volunteer moved over to sit with the man and orient him to where we were in the training.

After I answered a few more of the participants' questions on how to help others recover from trauma, the volunteer raised her hand and interrupted me.

"This man shared his story with me. Can he share it with the group?"

"Yes, absolutely," I responded, gesturing for him to take the floor. I always built in time for survivors to share as part of the training, so this was a natural occurrence.

Life hung hard on the thirtysomething man, so hard I wondered if this had been the case even prior to December 26. As he spoke, anger pierced his tone of voice and his eyes glared at the ground. I sensed he was a man set to blow, needing only a little push in the wrong direction.

I had plenty of exposure to volatile situations with my experiences in the military and hospitals, and I could see there was depth to this man's anger. Life had betrayed him, and he could not reconcile what had happened.

Because he was a survivor, he counted among the lucky. While he lost many family members, his wife and infant daughter survived with him. Beyond this seemingly hopeful situation, though, lay the ruins of his livelihood.

He lost his little home he had worked so hard to have. He lost every single personal possession except for the clothes on his back and his driver's license. He lost his ability to earn a living because he could not make more than the daily rent on the motorized pedal-cab taxi he drove. He and his family now lived in a camp for displaced persons with no idea what would happen next.

The staccato rhythm of his angry recounting lasted seven minutes. After he finished telling his story, the room lay silent. The eyes and faces of those surrounding him expressed deep caring. I hoped he saw this, as he made only intermittent glances up throughout his retelling.

"Thank you for sharing your story

with us," I said through a translator.

Hearing my comment, he stopped, looked straight at me, turned to face the door, and left.

Turning my attention from the door back to my audience, I knew exactly what to say next.

restoring hope

I had woken early that morning, laying in silence as the work team I shared my room with snored in the semi-melodic pattern resulting from a late night, hard work, and deep sleep. Prior to rising, I mentally reviewed the plan for the day, anticipation already building for the work at hand. Carefully making my way around the room, I dressed with minimal disruption to only the lightest of sleepers.

I emerged into the makeshift meal preparation area. Breakfast had been iffy for me since two days earlier when I came out to a breakfast of fried fish heads and rice. Now, I really have no problem with people eating fish heads; I enjoy a fish head on occasion myself. It's just not something I would usually put on my breakfast menu!

This particular morning I craved only a big bowl of Frosted Flakes and milk. Alas, this was not an option, so I made do with some rice and a granola bar I brought from home. Paired with a cup of strong Acehnese coffee, it was a decent start to the day. Breakfast, along with the anticipation of the day's main activity, carried me through my early morning grogginess to being ready to head out.

I was excited to conduct a training for relief workers on the *Restoring Hope* curriculum Nicola and I put together specifically for this purpose. We named the curriculum *Restoring Hope* because, quite simply, tragedy robs us of hope. No matter what our culture, religion, or environment is, we spend our lives building toward the future. Yet, the recent events ruined the hope of so many because they could not even see a future for themselves.

It was very gratifying to help equip these amazing and willing relief workers with additional knowledge and skills to encourage them in their work with survivors. As I gathered my materials for the training, I wondered who would show up.

To date, the trainings usually included the staff and volunteers of small- to medium-sized service organizations. Just before the beginning of each training, however, a few survivors would usually filter in. They would come because so many of them had heard the relief workers say they needed trauma

TRAUMA COUNSELING

From a professional standpoint, one of the interesting aspects of traveling to Aceh was conducting trauma counseling and trainings in a region where historically there was no mental healthcare system. In the United States, if there is a need for trained counselors, there are literally thousands available a short plane ride from any place in the country. In Aceh there were few trained counselors.

Additionally, in Aceh, thinking about one's mental health was either a luxury most could not afford or carried a stigma making it off limits, so there was very little knowledge infrastructure among the people on mental health-related issues to draw upon. Despite this, survivors started asking for "trauma counseling," not really even knowing what that meant, only understanding that what they experienced was so profound, it could not go unattended.

In this situation, my approach to working with relief workers and survivors alike was to focus on basic education and practical approaches to addressing trauma. I avoided any discussion of diagnoses. While an event such as the tsunami would understandably cause Post-traumatic Stress Disorder (PTSD), to the Acehnese, such a diagnosis only led them to believe they were "crazy."

This approach seemed to resonate with survivors and relief workers, providing direction on how to cope with trauma.

> "...OUR FOCUS WITH THIS WORK SHOULD BE TO REACH AT LEAST ONE PERSON, WHO PERHAPS DESPERATELY NEEDS TO HEAR ONE THING WE HAVE TO SAY."

NICOLA NOTE

Almost daily phone calls with Thomas allowed those of us at home to feel a part of his work in Aceh. His reports of trainings provided a strong connection to the personal experiences of the survivors.

counseling. More often than not, they didn't even know what trauma counseling was, but they trusted the workers, so they came to us whenever the signs were posted. Would any survivors come today? If so, I prayed for the words to encourage them and help them move forward.

I also remembered the words of my good friend Peter, who originally invited me to come to Aceh, from the night before.

"Remember, we're only here for the one."

I nodded in agreement. His words reminded me that our focus with this work should be to reach at least one person, who perhaps desperately needs to hear one thing we have to say. His words were grounding in the midst of so many serious needs.

the one

In a disaster area, everything becomes useable, so we held the training in the Restaurant Banda. We would conduct the training and then eat lunch together as a group when it was over.

As I rode to the training site, I mentally prepared myself for whatever was to come. The warm, humid morning air cooled with the acceleration of the car. As we wove through the sparsely trafficked streets, Peter's words from the night before to focus on "the one" comforted me. (He would be more right than even he knew.)

The Restaurant Banda lay centered on a street still showing the effects of the influx of water a month prior. While the ocean's long reach had not stripped everything to the earth, as was the case only a quarter mile closer to the sea, the damage was still evident. Spiderweb-like cracks in the façade of buildings ran the length and breadth of the structures. A mud line from the highest water level still marked most buildings.

The cleaning efforts of the owner of the Restaurant Banda invited customers to return, with a clean-swept front walk trying to mitigate the evident damage.

We walked into the restaurant. The stark white paint and fluorescent lights made us squint just slightly as our eyes adjusted to the flickering caused by the intermittent power supply. Peter and I began to set up as the staff and volunteer relief workers arrived and settled in.

Amongst the survivors who came, there was a pastor and a few of his associates. They represented a church we had visited whose flock had been decimated. A group of women, young and old, came after seeing a flyer posted on one of the surviving mosques. Four

fishermen came from a small village on the outskirts of Banda Aceh. More arrived until the room was full, and it seemed everyone who was meant to be there was there.

The training started at a quick pace and did not slow, with me speaking and Peter translating. The training area was set up with tables positioned in a circle. He and I stood in the center and did "training in the round." We fed off each other's energy as we communicated ways for each participant to help themselves and those around them deal with trauma.

It was energizing to train in a team format with a translator as capable as Peter. I saw recognition in the eyes of the relief workers and survivors alike, as light bulbs went off on how to apply this information. After thirty minutes of this rapid and vigorous exchange, the door opened and the angry man walked in.

Inevitably when I conducted these trainings, survivors would choose to tell their stories. I encouraged them to do so. I recognized the healing effect it could have on them and on those observing. As the man told his story, I looked for the ease that usually comes with sharing, but there was none.

After I thanked him for sharing and he walked out, I turned to my audience. Their faces registered great concern. What did this turn of events mean in the context of what they were learning? The room was silent for a moment as I determined how to use this situation constructively.

"We have to keep in mind this man's anger is perfectly normal under the circumstances," I said. Referencing back to the phases of grief we had reviewed earlier, I reminded them how anger was part of the natural progression from tragedy to acceptance—acceptance not requiring the survivors to be happy about what occurred, but only that they could move forward in life and not be held hostage by the tragedy.

"If you encounter someone in the future who looks and acts like this man

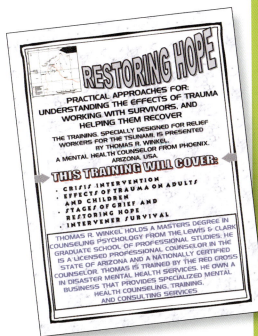

Restoring Hope

When traveling to Aceh, I had general materials relating to trauma and disasters with me, but no resources aimed specifically at helping survivors of a disaster of this magnitude or at training relief workers. And certainly none of the available materials were in Indonesian. It quickly became apparent that to conduct the trainings with the right information and resources, we would have to create a curriculum and training materials ourselves.

The *Restoring Hope* curriculum was born across phone and internet connections, and its creation spanned the globe. Over the phone, Nicola and I created a basic curriculum that covered the effects of trauma and how to help survivors. Then, she formatted it in the United States and emailed it to me in Aceh, where one of my fellow relief workers translated it into Indonesian.

We continued to make revisions and improvements to the curriculum thoughout my time there, making *Restoring Hope* current by the day. Later we had the opportunity to expand the curriculum even further when we conducted a three-day retreat for survivors several months later.

An Indonesian friend created this flyer to advertise Thomas' training based on the **Restoring Hope** *curriculum. His fellow relief workers were instrumental in coordinating with the small- and medium-sized relief organizations that desperately needed trauma counseling training for their staff, but didn't have the resources to provide training themselves.*

"I SMILED AT THE MAN... HE LOOKED BACK AT ME WITH A CHANGED EXPRESSION..."

Thomas poses with the daughter of "the one," a tsunami survivor in her own right.

did," I continued, "remember all they might need is for someone to not be afraid of their anger, but to listen to it. Despite the man's anger, he clearly felt safe enough to share his story with us. Whether this was the first, last, or hundredth telling, each one is significant."

"Many times we will never know the result of our interaction with a survivor. Every time, we can only put as much caring and compassion into our interactions as possible, and have faith we are having a positive effect, even though we may never see the result." The participants nodded, understanding this aspect of their work well.

About a half hour later, the door opened again. I was shocked as I saw the angry man walk through the door. He gently held the door open for a woman holding an infant dressed in pink. I noticed the participants look at each other with slight smiles of delight as they watched the man and his family enter the room.

I smiled at the man as he sat down. He looked back at me with a changed expression, so different from just a short while ago. Continuing with the training, I was so thankful for the encouragement I believed this encounter could be for the participants, reinforcing the powerful effect of

just an open ear and a warm heart.

As the training went on, the fishermen shared their great concern for friends back at the refugee camps who would only stare into space and smoke the occasional cigarette. The group discussed the challenges they faced, strategizing on how to help their friends.

The women shared their concerns for the children now joining new families of distant, little-known relatives or family friends. We discussed how they could band together as widowed mothers and share responsibilities, with some working while others cared for the children. We talked about all of these things and more, using every moment available. Finally, with the restaurant staff anxiously waiting nearby with lunch in hand, ready to place the plates on the table, we broke to share a meal together.

Quietly, the man who had held such anger shared with a group of us what had occurred since he left.

"I went to get my wife because I wanted her to hear what you were talking about so she could be helped too."

We smiled and told him we were glad he returned.

By the end of the meal, his anger had eased to the point that he was even able to smile. We took photos with his sweet baby as the conversation continued. Through the discussion, we determined the man had a skill in high demand in the torn and confused streets of Banda Aceh. Many of the non-governmental organizations needed skilled drivers, and with the one remaining material piece of his past, his driver's license, he could drive.

Two days later, Peter and I were just leaving a roadside café after having cups of fresh Sumatran coffee when the man from the training halted his taxi in front of us. He had come there to meet with Peter to set up his new position as a driver.

As he got out of his pedal-cab, I could see a renewed spirit of hope in his demeanor.

Coming up to us, right there in the middle of the street, he looked me straight in the eye, smiled, and hugged me. Smiling back at him, I said a silent prayer of thanks for "the one."

LESSONS FROM ACEH
trust

All emotions have their place...

As a therapist, you learn to use anything and everything a client does to help them reach their goals. Silence, anger, chaos, and even rejection are all useful in the therapeutic setting. This concept was very important in Aceh. It was not our place to dictate how a survivor should grieve or respond to tragedy. It was our role to meet them where they were (fear, anger, sadness) and help them move forward.

Telling one's story is an important part of healing...

This scene of survivors recounting the details of their experiences occurred over and over in Aceh. Instinctively, many of the survivors and those helping them seemed to understand that sharing aloud was one path through grief. For those of us there to help, we learned that simply listening is powerful. Sometimes, if that is all we have to offer, it is enough.

Look for one thing...

In addition to looking for "the one," the story of this man also reminds us to keep looking for the "one thing" in any situation. The one thing that provides a tiny ray of hope. In this case, the one thing was his driver's license, a path to rediscovering the feeling of having a purpose.

the life

Saturday, February 19, 2005

I looked into the faces of hope *today, the future of Aceh, and glimpsed a day when this place will no longer be defined by what nature's fury wrought. When the people,* forever changed*, can move forward. And I said* goodbye, for now*.*

Date:
February 19, 2005

Time:
Morning

Place:
Banda Aceh, Indonesia

Setting:
On one of his last days in Indonesia, Thomas experiences how hope endures in the midst of tragedy.

return

I sunk heavily into the seat for the return flight home, my body fully ready to surrender to sleep for the next eighteen hours. All around me, normalcy ruled as the other passengers stowed their bags and took their seats in preparation for the flight from Singapore to Los Angeles.

Everything seemed slightly surreal as my mind tried to process the events of the last couple of weeks.

I thought of the dozens of people who supported me financially, emotionally, and spiritually, making it possible for me to take this journey.

I thought of the incredible relief workers I met, who would stay on to serve survivors long after I departed Aceh.

I thought of the survivors, who trusted me with their deepest emotions and allowed me to walk alongside them for a short while.

I thought of the long road of recovery ahead for Aceh and Indonesia.

And I thought about what I would say when asked, "*So what was it like over in Aceh?*"

missing

"A baby was born last night!"
The nurse, a fellow relief worker, shared this amazing news with a look of pure joy on her face. We rarely heard good news in Banda Aceh during the weeks following the earthquake and tsunami, so we greeted her announcement with smiles all around. All of the relief workers agreed to go to the local hospital together to visit the new family.

Throughout the trainings I conducted in Medan and Banda Aceh, I had seen first-hand how worn thin the relief workers were by their environment. Constant grief and loss surrounded them. Earthquakes and aftershocks kept their footing unsure. Bodies, recovered daily, still lined the streets. It was one of my last days in Banda Aceh, and after everything I'd seen and heard, I was feeling a little worn myself. I looked forward to sharing an uplifting moment of hope.

We especially looked forward to a scene of joy within the walls of a hospital. Thus far, hospitals were only reminders of how much was lost. For every person saved and recovering in a bed, there were multiple stories of heartache for family members swept away by the waves.

We made our way to the hospital, walking toward the nondescript building with excitement. Pausing before entering the hospital, my eyes lingered on the wall before me and I stopped.

Stretched across the entire length of the hospital wall were signs, each one displaying the word *dicari* (di-char-ee), which means *looking for* in Indonesian.

Underneath that phrase was the face, name, and description of missing family members. Some signs had just one photograph and description, but far too many showed a couple, several very young children, or even an entire family. Some signs were black and white, old and faded by the elements, and hanging on by just a corner. Others looked professionally printed, the colors still bright, the sign still firmly affixed to the wall.

As I looked at that wall, I grieved for the hope on display, the hope the people who created and hung those signs held for finding their loved ones. Week by week, it had become increasingly clear how unlikely a happy ending would be in this tragic situation.

Standing there, I felt the weight of my experience thus far. The stories the survivors shared with me echoed in my head. The father who told me, "*I lost my wife and four children to the water and I have never found their bodies.*" The disbelieving mother who cried, "*My baby, only a few months old, was torn from my grip when I was in the tsunami.*" The priest who lamented, "*I did not do more to save those around me. I forgot that waves follow earthquakes. How could I forget?*" I grieved with them all.

Shaking off the memories for now, I continued past the grim reminder of hope and loss, looking forward to a more hopeful scene inside. Then, walking into the hospital, we heard the word. A baby had died. I never learned the reason, but saw the tiny bundle in the corner of the room. Wrapped in a blue blanket with a pattern of hearts and teddy bears, the small form served as another reminder of how closely death lingered even six weeks after the disaster.

I felt the urge to leave the hospital with its stories of sadness and the evidence of loss papering the front. But where would I go? Sadness and destruction were everywhere in Banda Aceh.

new life

Then, an amazing turn of events. An Indoncsian nurse dressed in a crisp white uniform approached us. She wore a white scarf covering her hair, as is traditional for Muslim Indonesian women. Her smile warmed my heart. Despite the tragedy surrounding her, and having to walk by the wall of missing

American Red Cross
Boosting Hope

The American Red Cross response to the tsunami began only hours after the disaster hit. In close coordination with local Red Cross societies in the affected countries, the relief effort eventually involved over 22,000 volunteers from around the world.

Individual, corporate, and foundation contributions for tsunami relief were the highest in the organization's 125-year history. With this support, the Red Cross implemented programs to address physical needs, such as housing, clean water, and food, as well as psychological support and livelihood programs throughout Southeast Asia.

In Indonesia in particular, the Red Cross has supported an extensive immunization program for children, including programs for measles, rubella, and polio. The polio campaign, in coordination with the Indonesian Red Cross, represented the largest immunization effort in Indonesian history with 23.4 million children vaccinated. Education and distribution of insecticide-treated nets are also helping to substantially decrease the incidence of malaria.

These efforts in health, combined with the other Red Cross initiatives in Indonesia, are helping ensure the future health of Aceh's children.

Source: "Tsunami Recovery Program - Strategic Plan 2006 - 2010" and "Two-Year Report," American Red Cross

"DESPITE THE TRAGEDY...SHE STILL RADIATED JOY IN SERVING OTHERS."

people every day, she still radiated joy in serving others. With her bright smile, she brought us into a room and showed us not one, but two newborn infants swaddled in matching blankets.

Laying side-by-side, only their tiny red faces peeked out, still shocked by the light and air. I noticed that the blankets had the same pattern of hearts and teddy bears as the other infant's swaddling.

These children, born to a survivor of the tsunami that day, showed all of us how hope endures in the midst of tragedy. One young life lost, but two lives started that day in the hospital. Gazing down on these twins, I felt renewed by such a strong reminder that life does indeed go on.

We took the opportunity to greet the family who gathered to celebrate the occasion.

"Congratulations," I said to them through a translator.

The exhausted mother only smiled, but her mother, the proud grandmother of the babies, told us, "The nurses made the babies live." In a time and place with very limited medical resources and many potential complications of childbirth, the nurses had ushered in new life against great odds.

Then my friend offered to take my photograph with three of the nurses working that day. In the photo, initially you see only our differences. I am a Christian, American man. They are Muslim, Indonesian women. My height towers over their small frames.

Looking closer, though, you see that what unifies us is our smiles. Initially brought together because of disaster, we now celebrated new life together.

Their smiles and joy remind me to this day that hope lives, even in the midst of tragedy.

home

So what was it like in Aceh?

The plane dipped low in preparation for landing at LAX.

It's hard to put into words.

I gathered my belongings for one last, short flight home, my mind and body on automatic pilot as I made my way from terminal to terminal.

The heart of the people is amazing... they are resilient.

Walking through the airport was surreal. How odd to be back in what should be my comfort zone, only to have it feel so strange.

It was life-changing.

I didn't know it at the time, but

a group of people, including my wife and several friends who made the trek over from Phoenix to surprise me, waited in the terminal to welcome me back stateside.

Then, I saw them. Faces I recognized. Friends. Home.

Really, it would take a whole book to describe the experience…

Thomas stands with some of the nurses from the hospital in Banda Aceh, sharing a moment of joy with the staff and family of the babies.

LESSONS FROM ACEH
choose joy

We get to choose our focus…

Surviving in a zone of chaos and trauma like Aceh required each of us to choose everyday to focus on how we could help. This kept us moving forward. The reality is that in life, we don't have to focus on the negative, but we often choose to. I think of the nurse from this story often, and remind myself to choose joy.

Look for the miracles…

Miracles, big and small, abound, even in the darkest places. We had to seek out this hopeful scene of a family celebrating new life because sometimes the miracles lay hidden amongst the struggles.

Take it with you…

Returning from Aceh, I brought home the pain and sorrow of the survivors I met and the trauma of the community I lived in for a short time. But I also made a conscious decision to bring home the feelings of hope I witnessed and was a part of because I knew the importance of sharing those experiences with everyone back home.

INTO THE HEART OF DISASTER:

Profiles of Service

As I was preparing for my trip to Aceh, I heard snippets on the news about efforts by others to help with the tsunami relief effort. One effort that stood out was a unique approach by my hometown of Phoenix, Arizona. The City of Phoenix formed the **Rising to Help** *partnership with Food for the Hungry, an international relief agency based in Phoenix. Together, they adopted the community of Meulaboh (muh-law-bow), Indonesia, with a ten-year commitment to help rebuild.*

This was a historic effort. As far as it is known, this was the first time a governmental entity and a non-governmental organization partnered to adopt a foreign city with such a substantial and long-term commitment.

Several months after my return from Indonesia, I would get to experience for myself how this partnership brought two very different communities together in friendship.

A Partnership of Hope

Photo courtesy of Food for the Hungry

the partnership

On January 7, 2005, City of Phoenix Mayor Phil Gordon, Phoenix City Councilwoman Peggy Bilsten, and Food for the Hungry President Ben Homan gathered to announce the historic partnership to adopt the community of Meulaboh, Indonesia. As of 2007, individuals, schools, and businesses have contributed over $350,000 to the effort, which is entirely funded by donations.

Photo by Rodney Rascona

first impressions

Peggy Bilsten and Ben Homan traveled to Meulaboh just weeks after the tsunami to meet with leaders from the city, assess needs, and lay out a plan for assistance. The leaders of Meulaboh expressed how grateful they were to have their authority, knowledge, and understanding of needs recognized, in the midst of a time when many foreigners were making decisions that affected their people.

Photo by Rodney Rascona

"As I walked through the neighborhood, the silence was deafening. You realize you're standing in the middle of a neighborhood that doesn't exist anymore...the families, the children that were there. I stood on a wood floor in what was a little girl's bedroom. Her bookbag was tossed, homework was ripped, and it was very hard to see a life that had been destroyed."

-Ben Homan
(excerpted from an audio report on the ground in Aceh)

the community

The city of Meulaboh, Indonesia, and the surrounding region, lost 40,000 of 120,000 residents in the tsunami.

Peggy's Journal - January 15, 2005

"The images I had been seeing on television, though graphic and horrible, simply cannot convey the enormity of destruction and human suffering that I am seeing first-hand. There simply aren't words to describe the smell, devastation, and loss of human life.

And the silence. It's a scary silence.

As we walked up and down neighborhood streets that should be filled with kids playing and families working, all that is left now are things like tiny tennis shoes, photo albums, bicycles, kids toys, and clothes scattered about.

There also are body bags, way too many body bags for me to comprehend.

Walking along another street, I witnessed one of the most heartwrenching things I've ever seen. There was what looked like a family—two adults and a child—still laying, decomposing in the middle of the neighborhood street. My heart goes out to all of the innocent people whose lives have been forever changed by this incredible disaster.

Families who lost everything are now living on the side of the road, in tents, and in water about two feet deep. And after all the suffering they've been through, when we walked by them, they still gave us a beautiful smile or wave, and said a quiet 'hello.'

With all the death and destruction, there still is life and compassion."

rebuilding lives

The focus of the City of Phoenix/Food for the Hungry programs is to create sustainable development in Meulaboh. Participants in programs like these have an active role in rebuilding their own lives and their community.

Cash-for-Work - This program provides an income for survivors, while also benefiting the entire community by repairing infrastructure around the city.

Livelihood - Grants and support provide the tools necessary for survivors to resume their trade, whether shears for a barber or a saw for a carpenter.

Agriculture - Rehabilitation of fields destroyed by saltwater and debris, as well as the provision of seeds and tools help Aceh farmers.

Education - After-school programs, English camps, and a teachers retreat help encourage staff and students.

Photo by Rodney Rascona

Photo by Rodney Rascona

Peggy's Journal - May 20, 2005

"I was honored to visit a school with Meulaboh's Director of Education. He is a wonderful man—passionate about children and education. His task right now is overwhelming. He told us that 164 of their schools were destroyed. Even more devastating, they are dealing with the deaths of 237 teachers, more than ten percent of their teaching staff, and more than 3,000 students. The education official wondered why we in Phoenix care so much. I told him we share his city's passion for young people.

These children have almost nothing. They use a soda can to play soccer. Their school has a playground about the size of two basketball courts. The area is all rocks and hard surface with no grass. All I could think of was how fortunate our children in Phoenix are to have something as basic as grass on their playgrounds.

Meulaboh residents are so thankful that others care. They love and trust the Food for the Hungry team that has been helping them for nearly five months.

That appreciation is probably what's most striking about my second trip to Meulaboh. What people here are saying is, 'We don't know why you love us so much, but we are so grateful.'

I am grateful to be here to witness the start of this long-term project. The message I will bring home is what a difference the outpouring of love from Phoenix has made in thousands of lives. It has given them hope for a future and provided warmth in their time of grief."

Photo by Rodney Rascona

friendship

A genuine friendship developed between representatives from Phoenix and Meulaboh. Councilwoman Bilsten interviews the *Camat* (cha-maht) or mayor of Meulaboh. Heidi Blomberg, who spent a year serving in Meulaboh, sits with the Camat's wife, Ibu Eva, in the ruins of a building destroyed by the tsunami.

Photo by Rodney Rascona

❝ *There is definitely renewed hope for the future as they know there are people who are going to stay around and help them as long as necessary.* ❞

— Heidi Blomberg

❝ *Food for the Hungry has connected the communities of Phoenix and Meulaboh in a way that not only promotes a good working partnership but also produces long-lasting friendship.* ❞

— George H. W. Bush

Photo courtesy of Food for the Hungry

Former President George H.W. Bush with Phoenix City Councilwoman Peggy Bilsten and Food for the Hungry's John Frick.

the team

Monday, September 19, 2005

Starting out as strangers, we are becoming a family. *We gathered to come and serve those who survived the tsunami against all odds.*

Into the midst of people who less than a year ago would not have believed our kind intentions, *but who now put their hearts and minds in our hands. And in return,* we are changed.

Date:

September 18, 2005

Time:

Morning

Place:

Lake Toba, Indonesia

Setting:

On very short notice, Thomas returns to Indonesia with Nicola and six other teammates to provide trauma counseling to survivors of the tsunami in the retreat setting of Lake Toba. They travel there via Medan, Indonesia, a familiar setting from Thomas' trip in February.

 back

I looked out the airplane window as the expanse of the city of Medan, Indonesia, came into view. Even at altitude, I felt the electricity of the city's heartbeat, a familiar vibe after the time I spent there seven months earlier. Landing at this very airport in February, I had no idea what lay before me, only that I was called to serve. The lost memories of the feeling of being a stranger in a strange land, all alone, washed over me.

But this time I was not alone. I glanced over at my wife, Nicola, as she peered through the window intently at the approaching land below. Even though she was with me in spirit every step of the way in February, when I returned from Indonesia, I vowed that I never wanted to tackle a mission again without her physically by my side. Now we were there to serve together, a team in every sense of the word.

Around us in the plane sat six other people. Despite thirty-four hours of travel, four flights, and a midnight to 7:00 a.m. layover at the Singapore airport, we were still relative strangers, though not for long. It is remarkable how close the bonds of strangers become when united by a common cause.

Collectively, our purpose in traveling so far together was to bring knowledge and support to those who had suffered and to help them begin to heal from the trauma of their experience. We were also there to document with stories, photographs, and video footage all that had happened to help ensure that the world does not forget.

Most importantly to Nicola and me, in addition to being our teammates, these six people also comprised our lay trauma counseling team, an essential component for the work we would do. With only one month to prepare a three-day trauma counseling curriculum, only one day to recover from jet lag, and only one hour to train them on everything they ever needed to know about trauma, this trip would prove that, with willing hearts and a united purpose, amazing things are possible.

divine timing

"Let's go to Scottsdale Bible for service this morning," Nicola said early one Sunday morning five weeks before. I groggily looked up at her through the morning haze of sleepiness, a little surprised at her suggestion. Scottsdale Bible Church is a large congregation that is a moderate distance from our house. While I occasionally attended service there during high school and the military, she and I had only attended a service there one time together. Though the suggestion seemed out of the blue, I said "sure," and we left to attend the 8:00 a.m. service.

It was August, six months since I returned from Aceh. I had spent those months focused on our counseling practice, helping clients who were so understanding when I was gone in February. Fairly constantly on our minds, though, was the question *what's next with Aceh*? We felt that our work there wasn't finished, and yet we didn't see any clear direction about what we should be doing next. What we did know is that when it was time to go back, we would do so together.

Running just a couple of minutes late, we missed the very first announcements and information of the service. Settling into our seats, I did a very unusual thing. Normally, I don't spend much time reading the church bulletin, but for some reason, this time I picked it up and leafed through it. Then a word popped out at me...*tsunami*. I leaned over to Nicola and pointed at a blurb that said the organization Food for the Hungry was in the courtyard providing a donor report on the funds the

congregation donated for tsunami relief. We decided to go over after the service and say hello.

To be honest, we felt a slight hesitancy. Over the past few months, we'd met several people who were involved in tsunami relief. Every time, we would get really excited. We wanted to talk all about it with them, compare notes about time and place, and just hear about what they had done. And almost every time, we were met with a lukewarm response. People didn't seem all that interested in talking about their experiences. We found this so odd given that unique experiences often serve as a connecting point, even among strangers. We prepared ourselves for a similar response, and expected a brief chat.

After the service, we walked up to the table. Standing there were Peggy Bilsten, a Phoenix City Council member, and John Frick, a representative of Food for the Hungry. The feeling from these two was definitely different from the start. They weren't just interested in Aceh; they were passionate about Aceh. We said hello and introduced ourselves, mentioning that we took a special interest in Indonesia because I went to Aceh in February. Chatting with John, Nicola offhandedly mentioned that I had done trauma

"...THIS TRIP WOULD PROVE THAT, WITH WILLING HEARTS AND A UNITED PURPOSE, AMAZING THINGS ARE POSSIBLE."

counseling. The response was unexpected…I thought John might jump out of his skin!

"We need to talk!" he said. He hurriedly explained that they had a trip planned to Indonesia the following month. The plan was to do a retreat for survivors, but the people who were supposed to do the trauma counseling portion were no longer able to go. Well, that was a pretty different response than we'd had anywhere else. He ended by asking me to pray about the possibility of joining the trip. I smiled to myself, thinking, *I already took care of that!*

We exchanged business cards and planned to talk that week about the possibilities. Walking to the parking lot, I couldn't stop smiling to myself, as I started to see more and more clearly the divine timing at work.

As we slid into a booth at our favorite breakfast place, Nicola and I looked at each other with "can you believe this?" expressions. Then I told her about what I was thinking just the night before. After six months of wondering what was next with Aceh, I felt an incredible conviction to do something more for the tsunami survivors. The strength of the conviction moved me deeply, and I prayed for direction. I fell asleep

telling myself that tomorrow, Sunday, Nicola and I would figure out what to do next. Sometimes our prayers are answered so clearly, there is no room left for doubt.

Over the course of the following week, Nicola and I agreed to join the Food for the Hungry/City of Phoenix team to provide trauma counseling at a three-day retreat for teachers in their education program. Less than five very short weeks after we made that fateful decision to go to Scottsdale Bible, we were on a plane to Aceh, Indonesia, no longer wondering, *what's next*? And this time, we were together in body as well as in spirit.

getting to know you

"In about two hours Thomas and Nicola are going to train us on what the team will do during the trauma counseling portion of the retreat,"

Photo by Rodney Rascona

John said before part of the team departed to do some filming in and around Medan.

After thirty-four hours of travel, just a couple of hours on the ground, and a delicious Indonesian lunch at the Food for the Hungry guesthouse in Medan, two hours didn't seem quite adequate for getting in the mindset to train. Especially when we needed to be alert enough to train the team on trauma counseling.

The jetlag was fierce, so Nicola went to catch a few winks while I reviewed our materials. I was hoping to keep my head clear enough to make sense. An hour later, I went to try to wake Nicola up; "try" being the operative word. Even though her mind knew it was time to get up, her body seemed quite unwilling to relinquish a state of sleep. Hoisting herself to a sitting position, she tried to pretend she was awake. But as I started to review the material for our training, she kept falling asleep every ten to twenty seconds, sometimes in the middle of talking!

I would start laughing, and she would rouse from her micro-nap

After thirty-four hours of travel, the team was all smiles as we landed in Medan, ready to begin the work at hand.

laughing with me. (Honestly, she could have fallen asleep a thousand times, I was just so happy to have her by my side.) After five laughable minutes of this, she finally splashed some water on her face, pulled herself into the living area of the house, and we finalized our discussion.

The team that had gone to film footage around Medan returned to the guesthouse. We all congregated in the living area. It was then that everyone got to hear the truly remarkable way our team came together. John described how each of us had an equally divine example of timing that led us to join the team. Meredith, a teacher, had only ten days notice of joining the trip, and she was getting married in just one month to boot!

When he was invited to join the trip, Pat, a Phoenix radio and television personality, told John that he had only two free weekends in the next six months. It just so happened they fell on the planned dates of the trip. Clearly, each of us was meant to be there.

Then, John turned things over to Nicola and me. We started by informing the team that we were going to condense my master's degree in counseling psychology,

THE PHOENIX TEAM

"the leader"

JOHN FRICK
Food for the Hungry

John's heart for the people of Meulaboh is inspiring. His ability to bring together teams to serve has made a significant impact on the city.

"the PR guru"

GORDON JAMES
Gordon C. James Public Relations

Gordon previously worked for the elder President Bush in the White House and toured the Maldives with former Presidents Bush and Clinton after the tsunami.

"the icon"

PAT MCMAHON
KAZ TV & KTAR FM

So dubbed by John in our first team briefing, Pat's status as "the icon" stems from his many years on television and radio in the Phoenix area.

"the teacher"

MEREDITH LEWIS
Alhambra School District

Meredith was our retreat counterpart, with her sessions focused on instructing the teachers on new skills for teaching their students English.

"the artist"

RODNEY RASCONA
Photographer

Before we even met Rodney, we were captivated by his ability to capture humanity on film.

"the writer"

ROSEANN MARCHESE
Food for the Hungry

It was Roseann's task to document the stories of survivors and capture the heart of the community of Meulaboh in words.

"the counselors"

THOMAS & NICOLA WINKEL
The Waypoint Group

Being a part of this team was an incredible opportunity to serve, and to work alongside and get to know our teammates.

Photos courtesy of Rodney Rascona

"...I HAD A SENSE THAT HEALING COULD TAKE PLACE IN THIS SPECIAL AND SERENE SETTING."

Lake Toba is a popular tourist destination in Indonesia, although the area has struggled since the Bali bombing in 2002, when tourism dropped off. This treasure in the heart of Sumatra is almost unimaginable in its secluded exquisiteness.

years of professional experience, and our *Restoring Hope* curriculum into the next hour of discussion. Fortunately, we received a few laughs. Adhering to the number one rule of afternoon trainings (let alone trainings on the tail end of circling half the globe), we passed out suckers to the group to hopefully help us all stay awake. A little bit of sugar, combined with the concerted effort of the team to stay alert, seemed to do the trick.

We reviewed the basics of our trauma counseling curriculum, and shared our approach to working with survivors. We briefly described the nature of trauma and some of the likely stories they would hear concerning death and disaster. "You are not expected to have answers for them," I said. "Just your support is enough."

We talked about cross-cultural issues and strategies for facilitating a small group. "Your role as small group facilitators is pivotal in this process," I said. I think this struck a cord of disbelief in some of them, but we knew that the intimacy of a small group often affords a level of sharing with great healing capacity. We also reminded them of the importance of taking care of ourselves as helpers.

We concluded the brief training with the assurance that the most important point for them to remember out of everything was that their caring is healing in and of itself. I looked forward to them seeing and knowing this was true.

toba

The boat moved swiftly across the water as we cruised toward Samosir Island. The island, which sits in the middle of Lake Toba (toe-bah), was formed by the eruption of a massive volcano 100,000 years ago. It was also the location for our three-day retreat with survivors of the tsunami.

From our vantage point in the boat, we started to get a sense of the beauty of the place we were heading to. A place where green mountains contrasted against blue water to create a striking and dramatic picture. On the bow of the boat, Rodney, the photographer, balanced his video camera to film Pat's take on the breathtaking landscape. I couldn't hear Pat's words, but I could see his animated expression.

My thoughts turned to our destination. Waiting on the other end of our short boat trip were twelve teachers, each of whom worked hard for the opportunity to attend this retreat. I wondered what their state of mind was at this point. Would they be open to the ideas we had to share? Would we be able to bridge the gaps of culture, language, and faith these many months after the disaster? Could we help them begin or continue the healing process from this tragedy?

I didn't know the answers to these questions, but somehow, as the boat approached the island, I had a sense that healing could take place in this special and serene setting.

I knew without a doubt that each of us was brought here to make a difference in at least one person's life. We started out as Food for the Hungry staff, trauma counselors, a professional photographer, a public relations expert, a teacher, and a Phoenix icon. A common purpose of helping those in need transformed us from strangers into a team.

Looking around the boat at our new friends, it occurred to me that they probably did not know the impact they would have on the survivors. In retrospect, it is clear to me that the willing hearts of these remarkable people made an impact that rippled out across Aceh.

LESSONS FROM ACEH
be prepared

Surrender the need to control your life...

For all the planning Nicola and I could have done to return to Indonesia, it is unlikely we could have created such a perfect opportunity to serve. We know we were meant to be on that team, to serve those survivors, and we know we could not possibly have orchestrated a better situation.

Ask for what you believe is best...

The initial invitation from Food for the Hungry was for me to join the September team alone. While this would have been workable, Nicola and I knew that it would be much more effective for both of us to be at the retreat. When we presented the benefits of us working together, an invitation was extended to Nicola as well. In the end, I know this was the best scenario for helping the survivors.

Equip people to serve and then let the heart do the rest...

Most of the people on our team had no experience with any type of counseling or trauma work. Our brief hour of training did not equip them with full knowledge about counseling, but instead provided them with an understanding of the nature of their role as helpers and supporters for the survivors. With this understanding, their humanity drove their interactions and resulted in incredible connections.

INTO THE HEART OF DISASTER:

Profiles of Service

Pat McMahon has been a familiar face and voice in Phoenix for decades. His antics on the long-running Wallace & Ladmo television show accompanied us both through childhood, and now as an award-winning talk show host, his voice is heard daily across Arizona.

He joined the Food for the Hungry Phoenix Team to see the devastation and reconstruction effort first-hand. We're fairly sure the impression he made on the people of Meulaboh well outlasted our visit.

Aceh! (God Bless You!)

by Pat McMahon

I remember someone calling the radio station asking if I wanted to go to Meulaboh. *Sure,* I thought, *I love exotic restaurants...*lunch or dinner? Ironic that I should ask that question considering the call was from a Phoenix-based international organization called Food for the Hungry. I had interviewed them many times on TV and radio about the remarkable aid they provide all over the world. I was impressed that they do so much more than feed those in need. Food for the Hungry recognizes all different kinds of emptiness...education...work...counseling, and they provide. So I said yes to the invitation to go as an observer to the tsunami devastation in the Aceh region of Indonesia. I was off to the city of Meulaboh.

But when I arrived, I was not prepared. I had covered earthquakes and floods, but I had never seen a hand come out of the sea and sweep away a large part of the population of a thriving city. Tidal waves aren't selective; they take parents and grandparents, boats and babies, schools and the students inside.

So why were the survivors smiling? What did they have to smile about when all of them lost somebody? When some of them were the only "somebodys" left in their families. Despite this, there seemed to be joy in their days, their work, and their insistence on a future. They truly are a hardy people.

I think the thing I will keep with me always is the change in their attitude toward the outside world. We had danced together, dined together, and, even without a common language, we shared many laughs. I asked one of the survivors whether we would have had a different relationship before the tsunami. She said that before the help came and before they got to know us, I might never have made it back to my room.

But we changed their minds...about America, about non-Muslims, about the West. We arrived, we unpacked, and we went to work without asking them to change. But because of the help, they did change, and so did their city. And so have I, forever, because I spent a week in Meulaboh.

Photos by Rodney Rascona

The United Nations is just one of the organizations that made a difference in the wake of the tsunami.

Now this is my kind of crowd! (I wonder how you say "Ladmo Bag" in Indonesian?) *Authors' Note: During the run of the Wallace & Ladmo show, a "Ladmo Bag" of toys and surprises was a coveted prize to any Phoenix child.*

Enjoying an afternoon with the survivors and relief workers on Lake Toba. (All I know is they said I would be going on a cruise with an outside cabin!?)

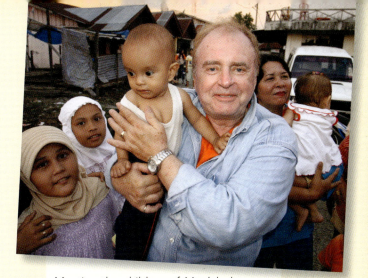

Meeting the children of Meulaboh...even after a disaster, I get to hold the future.

Sharing breakfast with two survivors during the retreat at Lake Toba. I'm trying to convince the Indonesian ladies I am Tom Cruise, the jet pilot in Top Gun. (It didn't work!)

On the peninsula on the outskirts of Meulaboh. Just reflecting on the things I've seen that I'll never forget and what we all can do to prevent people from suffering like this ever again.

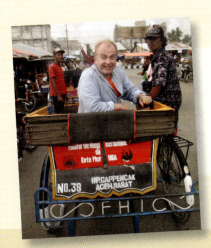

Riding around Meulaboh in a *becak* (bay-chak) pedal-cab donated by the City of Phoenix. (Hey, they got public transit from Phoenix before Phoenix did!)

the retreat

Thursday, September 22, 2005

What does it mean to move forward *after unimaginable loss? This group of survivors trusted us to take them to the depths of their grief, and in return, they are showing us the true nature of* resilience*. As we walk alongside them for a short time, we experience the beautiful mingling of the tears of grief and the* joy *of laughter in this special place of* healing*.*

Date:
September 20, 2005

Time:
Night

Place:
Lake Toba, Indonesia

Setting:
Thomas and Nicola, along with their teammates, provide trauma counseling to school teachers from Meulaboh during a three-day retreat in the tranquil setting of Lake Toba.

that's what it's all about

It is a year and a half later, and it is time to tell the story of the retreat. Out of all the stories, it seemed like it would be the easiest. In reality, it is the most difficult.

How do I share the experience of three days of grief and sorrow, while respecting the privacy of those who grieved? How do I convey the sense of destiny that permeated the retreat, each one of us meant to be there? How do I show the remarkable nature of the resilience of the human spirit in words?

Perhaps the only place to start is a room filled with a large circle of people. They came from around the globe, speaking different languages, holding different beliefs. And yet, through their movement and their voices, a unity emerged in purpose and in spirit.

"You put your left hand in, you put your left hand out, you put your left hand in and you shake it all about…"

It might have looked like a multicultural wedding reception gone awry or perhaps a cross-cultural musical exchange program. You would probably never have guessed that the people included survivors of an epic disaster and those who traveled around the world to help them.

Now, I have a confession to make. Under normal circumstances, I always sit out the "Hokey Pokey." The thing is, when a roomful of Acehnese teachers wants to learn to dance the American classic, you have an obligation to participate in showing them. So together, we put our whole selves in, dancing and laughing into the night.

On that September evening, at a retreat on the shores of Lake Toba, Indonesia, I made an exception and danced away with reckless abandon alongside people who had lost more than I could imagine. Looking back, I am still amazed at how, as a group, we traveled the road from grief to laughter and back again.

charting grief

Earlier that day, I stood at the front of a small conference room, with Nicola to one side and our translator to the other. We quietly discussed our next topic, an exercise on grief, as we waited for the teachers and staff to settle into rows of tables covered with bright green tablecloths. Along one side of the room, doors opened onto a covered patio, allowing for glimpses of Lake Toba and the lush green mountains beyond.

I glanced up as one teacher settled

into her seat. She had a kind and gregarious disposition, making it easy for anyone around her to smile along. Only later would we find out that her smiles and laughter masked a deep and powerful grief.

Another teacher found his chair. Through a "getting to know you" partner-sharing activity the previous night, Nicola learned that he spoke only a few words of English. He had a stoic, yet gentle demeanor, and I hoped that with the translator, the retreat would be a positive experience for him.

Our goals for this retreat were different from my work with survivors in February. These teachers were not in acute crisis. However, the disaster still affected them every day. Bethany Nanulaitta, the Food for the Hungry Education Coordinator who organized the retreat, described how some teachers still had a difficult time going to the third floor of a building, afraid they would not be able to escape if there was another earthquake.

In response to their ongoing struggle, we wanted to help the teachers think of a time before the tsunami without feeling overwhelming sadness for what they had lost. To stand on a beach without feeling immobilizing fear. To walk down the street without

constantly thinking of an escape plan should another earthquake suddenly hit. And, perhaps most importantly, to reclaim a sense of joyful connection to those they felt so horribly guilty for not being able to protect during that dark day. In essence, they needed to live their lives as normally as possible.

As we worked through the *Restoring Hope* curriculum, which Nicola and I had expanded from a three-hour training for relief workers to a three-day retreat curriculum for survivors, we wove together short lectures, worksheet activities, and small group time. We intentionally emphasized small group work because when the retreat ended and the team departed, the teachers would be each other's support system, and we wanted to model that concept.

I asked the group to turn to the page in their book titled *Kesedihan* (kez-eh-dee-hon), Indonesian for *grief*. I started talking about the five phases of grief—denial, anger, bargaining, depression, and acceptance—pausing periodically for the interpreter to translate. "Not everyone will experience the phases in the same order, and they can often overlap. The phases are not necessarily something to get out of... we want to move through the phases toward acceptance, so that we can feel

better," I explained.

On their worksheet was a large, blank circle. We asked the group to make a pie chart showing what percentage of the time they had spent in each phase since the tsunami (only we called it a pizza chart because they don't really have pie there, but they know about pizza—an example of cross-cultural adaptation!). The exercise encouraged their awareness about their own individual grief process, and also provided

We started our first session by presenting small gifts to our participants, including postcards of our home city of Phoenix, signed by our eight team members. These cards became a great conversation point between the team and the teachers.

Relief Spotlight

Mission Aviation Fellowship
Hope Flies

With their ability to fly into areas of Aceh cut off by damaged roads and infrastructure, Mission Aviation Fellowship (MAF) was onsite in Meulaboh almost immediately after the tsunami. Within the first two months, MAF conducted 1,114 flights transporting 387,743 pounds of food. Since then, they have supported the efforts of over sixty humanitarian organizations throughout Aceh through their aviation, communications, and technology services.

Tim Chase, a pilot with MAF, lived and served in Aceh with his wife, Karen, for two years. He shared, *"It has been an amazing privilege to be able to fly for and work alongside such a broad community of relief and development workers from around the world in support of the people of this region. In the face of such overwhelming need, we've played only a small part, but over and over again, people have expressed their appreciation for MAF's safe and reliable flight service...it has been the hardest we've ever worked, but it has also been the most fulfilling."*

Tim was one of the pilots who flew our team, staff, and teachers from Meulaboh to the retreat, turning what would have been a rough eighteen-hour bus ride each way into a one-hour flight.

Source: "Sumatra Program 2007 Overview," Mission Aviation Fellowship

context to help them make sense of the emotions they were experiencing.

This seemed like a relatively innocuous activity, but it struck a chord with the smiling teacher in particular. Pat McMahon, one of our team members, sat next to her. He described how the two of them had connected during the retreat, despite the language barrier. They would frequently laugh and joke around between sessions and at meals.

He recounted how she looked down at her grief chart to put pen to paper. In an instant, her smiles and joy dissolved into pain and grief. Through her tears, she shared with him the loss of her young child in the tsunami. As he and the others at their table listened to her, the official death tolls, so widely reported on the news, gave way to the individual human and heartbreaking grief of a mother.

serenade

That evening, before the "Hokey Pokey" and a group effort at "Row, Row, Row Your Boat" (in rounds no less!), the teachers took turns serenading us with traditional Acehnese songs. Their beautiful voices sang out clearly, intermixed with playful laughter. As I watched them, I marveled at the element of the human spirit that allows for laughter and joy, despite great loss.

At the end of our first full day of the retreat, we all knew this experience was something very special. The teachers, who all worked hard to earn a spot in the retreat, accepted the information we had to share, as well as the support and encouragement of the team. They were fully engaged in the process, whether getting to know members of our team or completing an exercise on trauma and grief.

The members of our team and the local staff were seeing first-hand just how far basic support and listening could go in helping those recovering from trauma. Several commented that they did not anticipate how much it would impact them personally to participate in this process.

Villagers on Samosir Island in Lake Toba gather at the beach to get an up-close look at the Mission Aviation Fellowship seaplane.

The coordinators of the retreat were seeing the positive effect of their decision to provide a place for healing away from the disaster zone.

Nicola and I were seeing the great worth of lost sleep from late nights spent adapting our curriculum for the retreat setting.

The group was bonded together by shared experience, and trust was built between those who suffered and those who came to help.

With this, I knew that all of us were ready for what was next.

guilt

The following morning, Nicola and I rose before the sun. Sitting outside, we waited for dawn to break, talking quietly about the plan for the day.

This day would be different for the survivors. This day, we would try to help them come to grips with their survival and their guilt. With one in three people in their community lost in the disaster, they needed to believe there was a purpose for their survival.

Survivor's guilt is a natural, yet destructive, response to trauma. It makes people believe they should not have survived, when so many others perished. It tells them that if they were

smarter, better, faster, or stronger, the outcome could have been different for those they lost. It neglects the reality of the experience, that no one could be expected to think clearly and react calmly in the midst of the chaos and crisis of disaster.

As challenging as I knew this would be, I felt the teachers were ready, as each had already shown a great willingness to face their trauma so they could begin to move forward.

After eating breakfast together, we gathered in the training room.

"Today we will be talking about survivor's guilt," I began. As I spoke about the concept, I saw slight nods of acknowledgment.

We broke into three small groups and moved outside to the covered patio. Each group drew in close, both facilitators and survivors leaning in as each person shared their experience with survivor's guilt. The blue waters of Lake Toba formed a spectacular backdrop for a most difficult subject.

I walked up to a group as one teacher began sharing her story. The woman's expression told the story of her trauma.

She spoke of the wave coming, dark and black. How the current, strong enough to rip new buildings from firm foundations and aged trees from deep ground, carried

The retreat combined teaching skills and trauma counseling to encourage the teachers in the healing process and increase their ability to help their students.

THE IMPETUS FOR THE RETREAT:
To provide hope and encouragement

The Director of Education in Meulaboh asked if Food for the Hungry could do something to encourage his teachers. Additionally, nearly every city leader cited trauma counseling as a need. The retreat was a response to these identified needs.

THE TEACHERS: *Trying to move forward*

After the disaster, many of the teachers found it difficult to be motivated to teach and go about everyday life. Despite this, each of them completed a set of requirements to attend the retreat, including taking on the responsibility to share what they learned with their colleagues.

THE STUDENTS: *Struggling to learn*

Many of the teachers observed a lack of motivation and focus among their students after the tsunami. Test results in the months after the disaster showed only 36% of the students successfully passing with only a 42.5% pass/fail mark.

Our work with the teachers was built upon the existing relationship the Food for the Hungry staff had formed with the teachers. Because the teachers trusted the staff, who brought us in, they trusted us as well.

Perspectives

Bethany Nanulaitta

Bethany Nanulaitta spent most of the first seventeen years of her life in Indonesia. At the time of the tsunami, she was back living in the United States, but felt the call to serve.

"Immediately after the tsunami I was compelled to return to Indonesia. I wanted to help those that had lost so much rebuild their lives.

I continue to marvel at the strength of the human spirit. How is it that the Acehnese can smile and laugh and face another day after their entire lives were swept away. How can they pick up and move on when I struggle to do that with much smaller problems? God is using their lives to teach me gratitude for the things I have.

Aceh is open to the world. The 'porch to Mekah (Mecca)' is no longer isolated. The people are learning that outsiders might think differently or even dress strangely, but that they are not all bad. What a privilege to be one of the people demonstrating this in a tangible way."

Bethany used both her fluency in Indonesian and her education background to serve as Education Coordinator for Food for the Hungry in Meulaboh.

her small daughter away from her side.

"I yelled out, 'Help me! I can't find my child. Help me…I want to see my child! Where are you? Come back to me…'

But they never found my child.

I feel so guilty…why didn't I hold on to her hand tighter? Sometimes I still hear the voice of my child calling my name.

I am so sad…so very sad. I will never get to hug her again."

Those seated around her, fellow survivors, relief workers, and new friends from across the ocean, could not bring her daughter back. They could not take away her pain. They could not make this right.

Instead, they did what they could do. They sat with her. They listened to her painful story. They cried tears for her loss. And they lifted her up with the empathy etched on their faces.

Later, Gordon James, who was a part of that group, wrote, *"some people might be able to keep the tears out of their eyes while a parent tells about her four-year-old daughter slipping out of her grasp and disappearing into the floodwaters. I'm not one of them. Painful as it sounds, recalling these experiences and making peace with them is how the Winkels help people heal."*

In the end, guiding them back to tragedy allowed them to hope again.

conclusion

We walked along the shore of the lake, our path lit only by intermittent lamplight. Breathing in the clean air and breathing out the relief that comes from completing something truly challenging, I felt a deep sense of gratitude for our experience. We would be leaving in the morning, and a hint of sadness permeated the atmosphere. It was difficult to imagine when we might be back to this very remote region again.

The following day, we were heading straight into the disaster zone of Meulaboh, the closest major city to the epicenter of the earthquake. For me, it meant a return to a challenging environment. For Nicola, it would be her first foray into the true nature of disaster. It was also the home of the twelve teachers with whom we spent the last three days.

As we approached the terrace where our group congregated for one last evening together, my thoughts traveled back. To the compassion of our team, who, with only a brief training and their humanity, reached out so incredibly to those in need. To the willingness of the survivors to accept the love and support of strangers from a foreign culture who don't even speak their language. To the stoicism of the male teacher I wasn't sure if we reached. To the profound grief of the mother that brought us all to our knees.

We entered the circle and those very teachers greeted us with smiles and welcoming gestures. After a few songs and much laughter, Bethany handed out certificates to the teachers for their participation in the retreat. Each one came forward and shook our hands or hugged us.

At that time, the teacher who had been so reserved and quiet all week came up to us. Despite his very limited English, he leaned toward us and said perfectly clearly, "You are my teachers, and I will never forget you." Bowing our heads slightly forward in respect, we shook his hand and thanked him, knowing the same was true for us.

That last night with the teachers was not about counseling or trauma. It was simply about recognizing progress on the journey of recovery and celebrating new friendships forged across many barriers.

And *that's what it's all about.*

LESSONS FROM ACEH
find restoration

Get away...

Retreats are purposely held in peaceful places. We all need time to regroup before moving forward, and the mind needs space to relax, away from the constant intrusions of noise, obligation, and busyness. While it may not be possible to get to a place as secluded as Lake Toba, seek out a peaceful place close to home when you need restoration.

Reconnect to reality...

For many of the survivors I met, an exploration of guilt revealed an undue burden placed upon them. One mother felt guilty because the week before the tsunami she was trying to foster her young daughter's independence by having her sleep in her own bed, against her daughter's wishes. After her daughter died in the disaster, she believed she had a been a bad mother. Reframing her actions in the light of a concerned parent helped her reconnect to the child she lost and eased her unnecessary guilt. While this could not erase her pain, it released the part that was not grounded in reality, thereby allowing her to start moving forward.

You never know your impact...

Throughout the retreat, I could not have told you whether the quiet teacher was benefiting from the retreat. He seemed happy enough to be there, but it was difficult to gauge anything further. When he said, "*I will never forget you,*" he reminded us that we never know what impact we make on other people. Because of this, we must be vigilant about what we put out into the world with our actions and words.

INTO THE HEART OF DISASTER:

Profiles of Service

After the tsunami, people from all over the world descended on Aceh to help. While some stayed for only a brief time, many made the commitment to live and work in the community on a long-term basis as part of the relief effort.

Two such people were Mark & Heidi Blomberg. While they came to Aceh separately (Heidi with Food for the Hungry and Mark with PACTEC), living and working in Meulaboh brought them together. When they returned to the United States, they married.

Here, Mark & Heidi provide a glimpse into what it was like living, working, and playing in Meulaboh.

Photos courtesy of Mark and Heidi Blomberg

A Year in Aceh

by Mark & Heidi Blomberg

Heidi Blomberg

After spending three years in Peru and traveling extensively to places like Africa and the Dominican Republic, Heidi was based in Phoenix at the time of the tsunami, working for Food for the Hungry's Child Sponsorship Program. She immediately told Food for the Hungry she was willing to go to the tsunami relief zone. Just a few weeks later,

with only five days' notice, she was on a plane heading for Jakarta, Indonesia. Eventually, she made her way to Meulaboh, where she served for the next year.

Mark Blomberg

Mark grew up as part of a missionary family serving with Mission Aviation Fellowship (MAF). His childhood, through junior high, was spent in Africa, including terms in Zimbabwe, Lesotho, and Congo. At the time of the tsunami, he and his brother, Jonathan, were in the process of getting their airplane maintenance licenses in Idaho. In May 2005, when MAF and PACTEC (Partners in Technology International) needed staff to assist in Meulaboh for a period of several months, Mark and Jonathan finished up their program early and departed for Aceh within two weeks.

view from the sky

Heidi: One of the only surviving structures on this peninsula is the green-roofed Army barracks. At the time of the tsunami, the soldiers were standing in formation. All of them died when the waves came crashing over the building.

view from the ground

Heidi: It looked like a wrecking ball came through Meulaboh. In many areas, neighborhoods were flattened and nothing remained. I couldn't even envision what had been there.

heidi's first house

Heidi: One of our first tasks was finding a place to live and work in Meulaboh. Our first house didn't have running water, so I learned how to throw a bucket into a well, which takes more skill than you might think (make sure to hold on to the end of the rope!). The view from the back of the house was of a debris field.

making do

Heidi: Pete Howard, Food for the Hungry's Relief Coordinator in Meulaboh, sits at his "desk" made out of water bottles.

adjustment

Heidi: I expected to be overwhelmed by the chaos and destruction when I arrived in Meulaboh. When we landed, we went straight into work mode, and I actually didn't cry for months. In retrospect, this was protective because if I had felt all those emotions, there's no way I could have functioned and worked effectively.

Mark: By the time I arrived in May, the recovery phase was well under way. While the cause of the trauma was different, characteristics of the recovering community were similar to other places I had lived and seen. People were adapting and functioning the best they could.

collaboration

Heidi: We started with an informal assessment, asking a lot of questions about what was needed. We went to the office of the Camat, the leader of Meulaboh. We explained who we were and about Food for the Hungry, and asked what we could do to help. The Camat took us on a tour of areas where gaps existed in aid, helping determine where our efforts would be best focused.

mark's "house"— united nations base camp in meulaboh

Mark: My brother and I lived in the UN Base Camp, in a tent, for five months. Growing up in Africa gave our living situation a sense of familiarity. We immediately felt at home in this setting, even though we knew nothing about Indonesian culture when we arrived.

mark's work

Mark: Jonathan and I ran the PACTEC internet cafe at the UN Base Camp, which provided internet access for all the NGOs (non-governmental organizations) in Meulaboh. We operated the cafe and provided tech support. It was a different role in that we were there to serve the other relief organizations. I really enjoyed getting to know the relief workers, both Nationals and those from overseas.

Heidi: For the first three months, we worked from 5:00 or 6:00 a.m. through to midnight everyday, with evenings spent at the internet cafe communicating with the home office in the United States.

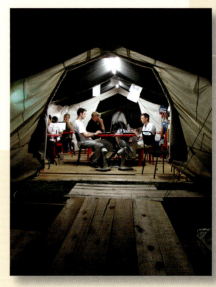

heidi's work

Heidi: One of the first programs we started was Cash-for-Work, where we employed local people to help clean up the debris around the city. Later, we started a program to provide grants to help people get back to work and become self-sufficient. This led to some interesting budgeting quandaries, such as: how many chickens should we buy (and how much does a chicken cost anyway?) and what do we feed them? Eventually, we helped people regain their livelihood in diverse areas such as agriculture, carpentry, baking, and barbering.

the beach

Heidi: One of the ways I learned to take care of myself was by running on the beach. Ironically, the ocean, which wrought so much tragedy, was the one place I could feel peaceful in Meulaboh. The ocean to me was an expression of God's power. On Sunday afternoons, we all headed out to the beach together. We would grill shrimp, tuna, red snapper, or whatever we had. We swam, played frisbee, and enjoyed the beautiful sunsets. It was truly a relaxing time. Our other activity for relief from work was escaping into TV shows like *Seinfeld*, *24*, and *Lost*, which we played on our computers.

the people

Heidi: I loved the people of Meulaboh. They are wonderful, beautiful people. They were kind and generous, inviting us into their homes. Some would ask, *"Why did you come? You're not Muslim, you're not Indonesian...you left your family to come here."* I would reply, *"We love and serve God just like you do. It is our responsibility to help others."* This was a connecting point between Islam and Christianity.

Postscript: On the surface, a relief community in the midst of a disaster zone might seem like an unlikely place for finding one's life partner. It somewhat makes sense, however, when you consider that the people you are around are like-minded in their life focus. Heidi and Mark, who married in October 2006, are one of three couples we know of who met in Meulaboh and became engaged.

the city

Saturday, September 24, 2005

This is a place of *contrast*. Brick by brick, a new home is built, right alongside the ruins of one claimed by the waves. At times, *tears* roll down the faces of survivors, but later, hopeful smiles and *laughter* emerge. We watch amazed as a city and its people *rise from the rubble*.

Date:
September 23, 2005

Time:
Late Morning

Place:
Meulaboh, Indonesia

Setting:
Thomas and Nicola arrive in Meulaboh, the closest city to the epicenter of the earthquake.

Photo by Mark Blomberg

survivor

The small plane cruised smoothly through the clouds, the lush green land of Sumatra laid out below. Lake Toba and her secluded beauty were soon behind us. Ahead of us lay Meulaboh, a wounded city now recovering.

As the plane approached the airstrip along the coast, Nicola pointed out tiny squares dotting a thin ribbon of a road. Some squares were light grey, the foundations all that remained of what had been people's homes. Other squares were bright blue, the tarps stretched across them serving as shelters nine months after the disaster.

My eyes lit out to the vastness of the sea, spreading endlessly to the horizon. For just a moment, from my vantage point in the sky, I pictured it rising up in anger and crashing upon the land. It hit me once again…*every survivor is a miracle*.

Only later did I realize there was one story I had not fully considered. One perspective unique to all others.

What story did the city have to tell? What trauma threaded through its veins of streets and walkways? What fear permeated its soil? What hope did it now feel? What would it tell us of that day and the days since, if it had the words?

the city speaks

"It was the darkest of all days, but it did not start out that way.

On the surface, nothing was amiss that morning. As the day began to unfold, no harbinger of warning arrived. And so the people went about their day, unaware of what stirred beneath the sea.

They had no warning, but I did.

I knew something was very wrong. An uneasy tension built deep below. I am no stranger to this tension; it is part of existence here. Great tension is constantly counteracted by quiet rumblings, a delicate balance maintained.

This time it was different, unlike anything I had felt before. No gentle rumblings relieved the growing tension. It built and built unrestrained, and as it did, I was afraid.

Afraid for the fisherman, bobbing on the sea. Afraid for the young mother feeding breakfast to her small children. Afraid for the woman bent over her washing. Afraid for the man winding his motorcycle through the streets.

And what would I tell them if I could speak? 'Run?' Where could they go? Where would they be safe?

So I could only wait.

Then it happened. A deep and terrible noise arose from far below the ground, the sound of tension released. And suddenly, nothing was as it should be. Relaxed conversation turned to screams of terror. The very ground

124

transformed into an undulating surface where no footing could be found.

As the ground shook violently, I tried with all my might to be strong, to protect them. But I could not prevent the buildings from collapsing. I could not prevent the damage to all their worldly possessions. I could not do anything.

Finally, thankfully, the shaking passed through and the ground ceased moving. And it was silent.

Fear gripped me as I waited in the silence, steeling myself for more silence.

Then, small movements as they emerged from homes, buildings, and cars.

Alive!

Though some were lost and damage was great, most survived. We survived. Relief washed over everything and everyone.

Many sought to find their loved ones, to make sure they were unharmed. Others just collapsed to the ground, so grateful to be alive.

Then, a new nightmare. The peaceful coexistence between me and the sea shattered. I sheltered them and the sea provided for them, but that day, the sea rose up against us.

It roared and came dark and black. I heard screams of terror. I felt footsteps, running in vain to escape.

Worst of all, I could offer no comfort or protection to them, for I was overtaken, screaming out my own fear. Drowning.

It seemed endless, but finally it subsided. As the sea retreated, I was left in ruins.

relief

Those first days were the darkest.

We had always lived in relative isolation, content among ourselves. Now it seemed this isolation might mean the end of us.

So many were lost, and I knew more would follow, but I could offer little as my gashes and wounds lay exposed to the elements. What would become of us?

Some cried out in despair. Others were eerily silent, the trauma clinging to them, stifling their expression.

Then another sound; the sound of the outside world. Helicopters descended, landing amidst my ruins and the vast crowds of the hungry. As they passed out what little they had in the way of goods, they soothed the worries of many with small gestures of kindness.

It was disconcerting at first to have outsiders among us, but there was kindness in their approach, and slowly trust built.

They came to feed the hungry. They came to heal the sick. They came to shelter the homeless. Speaking in words I did not recognize, but understood just the same, they spoke of comfort, compassion, and hope.

And then, they began to heal me. With debris cleared and roadways restored, life returned.

But how long would they stay? The answer came…we will walk alongside you.

So they stay, long past the time when our plight is illuminated, and yet the help is still so needed.

"IT HIT ME ONCE AGAIN… EVERY SURVIVOR IS A MIRACLE."

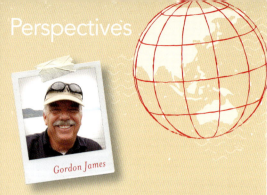

Gordon James

Gordon James accompanied former Presidents Bush and Clinton to the Maldives after the tsunami. He shares his perspective on the devastation and hope in Meulaboh nine months after the disaster.

"My trip to the Maldives did not prepare me for the devastation in Meulaboh. There is still standing water. People are living in camps designed to last just six months. A mass grave brings home the loss of life. But one thing struck me harder than the devastation: the continuing optimism of Meulaboh's people.

The City of Phoenix has left its mark in many parts of Meulaboh. One man, who received a pedal-cab to start his own small business, has a sign on his cart thanking Food for the Hungry and Phoenix. That sums up so much of the Meulaboh experience: Destruction and sadness pushed back by hope and new beginnings.

Rebuilding Indonesia is going to take a long time. Though we have so many of our own to look after, we cannot forget our friends on the other side of the world. To the people of Meulaboh, Phoenix is a city of caring people that wants to help them rebuild, a place where good deeds don't know international boundaries. I will not soon forget the people I met here, nor their great spirit."

Gordon is founder of Gordon C. James Public Relations.
Photo by Rodney Rascona

I have always had a connection to the people that choose to live here. Each one is a part of me.

Remembering what it was like before that dark day is difficult. That time seems so distant, and yet, occasionally, I have glimpses of how it used to be. When laughter permeates the hearts of the children once more. Or when a young mother makes plans for the future. Or when a fisherman once again goes out to the sea. Then I remember what it was like. And then I believe it can be like that again."

a city rising

As we drove from the airport to the center of Meulaboh, I could feel the burgeoning hope of the city as its physical wounds continued to heal.

Now, my attention turned to the emotional wounds of a new group of survivors. That afternoon, Nicola and I would provide training for the leaders of the local women's organization, the PKK (*Pemberdayaan Kesejahteraan Keluarga*, which roughly translates as Family Welfare Empowerment).

I said a silent prayer for focus and renewed energy. The intense heat and humidity of Meulaboh, such a dramatic change from the coolness of Lake Toba, hit me hard as soon as we landed. Driving through the battered streets of the city brought back into sharp focus the experiences of the survivors we had worked with all week. The intensity of facilitating the teachers retreat left Nicola and me exhausted. But soon, a new group of survivors would sit before us, and we needed to be fully present.

That afternoon, we rode in pedal-cabs to our training site where we met up with Heidi, who coordinated the women's training. We carefully climbed the steep staircase to the top floor of the three-story building, emerging into a sun-filled room. As we surveyed the space, discussing the format for this presentation of our *Restoring Hope* curriculum, I felt the tiled floor press uncomfortably against my bare feet (it is a custom in many homes and offices in Indonesia to remove your shoes at the door). As a presenter, you don't realize how much you rely on a certain comfort zone, until something as basic as shoes are taken away!

"Leaders from the twenty-one villages are coming for the first training today," Heidi explained. "For tomorrow's larger training, they will bring the women from their local groups who are struggling the most with trauma."

Heidi then introduced us to Ibu Eva Mahnizar, the wife of the Camat (cha-maht). Heidi explained that as the wife of the Camat (the equivalent of the mayor of Meulaboh), Ibu Eva was the leader of the PKK women's group. It was immediately apparent to Nicola and me that her strong, yet gentle presence made her a respected member of her community. We appreciated her leadership in organizing trauma counseling training for her people.

It was our first chance to meet Ibu Eva in person, but already her story had touched us. While she and the Camat survived the earthquake and tsunami, they suffered great loss. Both of their mothers were lost, as well as more extended family, including the wife of the Camat's brother.

In the aftermath of the disaster, it fell on them to provide leadership to their people despite their own trauma. Coordinating closely with non-governmental organizations, they worked tirelessly to rebuild their city and the lives of their people.

As the months went by, a mutual respect and friendship grew between the relief workers and the Camat and Ibu Eva. This bond was strengthened through months of working together.

However, the tsunami had not claimed its last victim. Five months after the disaster, the Camat's brother, grief-stricken over the loss of his wife, fell into a diabetic coma and passed away, leaving behind three young children.

Pete Howard, the Food for the Hungry Relief Coordinator in Indonesia at the time, talked about the loss of the Camat's brother:

"When Heidi and I arrived at the Camat's home with a condolence gift, we were met by the traditional mourning scene in the Islamic world of family and friends visiting. When the Camat and Ibu Eva saw us, they hugged us and invited us into their home for tea, pouring out their hearts with questions and sadness. What they couldn't communicate through their words was amply expressed by the sorrow and confusion in their eyes. He talked about how he didn't understand the loss...so soon after losing so many other family members to the tsunami. He then said how important friendship was to him and he marveled at the kindness that we had shown to him and his wife, saying 'You have shown greater care and compassion than anyone...'

He went on to say he doesn't understand God's ways, but thought maybe God was answering, through the pain, the prayers we had prayed with him and his wife to have children. 'Maybe we have our answer from God to have children...we are now parents overnight to three children,' he said. We told him that we would be praying for him and his wife as they become parents and as they tried to understand their loss once again. When we said this, he buried his head in his hands and then looked up and said simply, 'Thank you!' We then told him we would pray that 'the God

Indo Insights

Getting Around

Sometimes traveling to a foreign country is a great reminder that, no matter how chaotic driving in the United States might be, it positively pales in comparison to most of the world. The rules of the road in Indonesia generally seem to be, drive where you can find space (we even saw a car driving on the sidewalk in Medan), and honk your horn often so other drivers know you're there. Add numerous motorcycles weaving in and out and you end up with organized chaos.

Another common method of transportation is the *becak* (bay-chack), a small pedal-cab powered either by a bicycle or a motorcycle. One of the Food for the Hungry programs provided one hundred becaks to drivers in Meulaboh. Previously, these drivers had to rent their becak, but now they are small business owners. It was wonderful to see the symbol of the City of Phoenix, the firebird, painted on the back of the gifted becaks throughout Meulaboh.

Photo by Rodney Rascona

Evie Nirwana

Evie Nirwana served as a livelihood coordinator with Food for the Hungry in Meulaboh. She reflects on the changes in Aceh since the tsunami.

"Before the tsunami, the people of Aceh were closed to the outside world and there were very few Christians in the area. I did not know how the people of Aceh would receive and respect me as a female leader of a different religion in my position as a livelihood coordinator.

I was touched by a particular experience of a father who lost his wife, two children, and his livelihood. With tears in his eyes he said to us, 'Those who came to help when I am in need, I accept as my family. Three days after the tsunami, we had no food, but the NGOs [non-governmental organizations] gave us food, they gave us clothing, a place to sleep, and also a stove for cooking. Now that you are coming to us to help us rebuild our business, you are my family.'

After working in Aceh I realize that the most important thing is love. Love can change people's hearts, and love can overcome our differences."

of all comfort' would comfort them in their sorrow (2 Cor 1:3 NIV)."

Out of this tragic story came a ray of hope. We learned that the Camat and Ibu Eva, married for ten years, had not been able to have a child, though they wanted a family very much. They adopted his brother's children, creating a new family. Children no longer parentless. Parents no longer childless. Something new and beautiful rising out of the rubble.

As the women of the PKK filed in, I felt the familiar combination of heavy responsibility for helping those who lost so much and profound gratitude for being equipped to do so. Preparing to begin the training, I felt fatigue leave and focus return, ready once again for the work ahead, even in bare feet.

ripples

That evening, we toured Meulaboh. From the ground, we had a chance to see more of the devastation, but also more of the vibrancy of life returning to the city. Our trip ended on the

> **NICOLA NOTE**
>
> *It was wonderful to see the strength of the women of the PKK, who are so focused on the well-being of their community.*

peninsula on the outskirts of the city. Here, the enormous waves hit from two sides, and what was once a residential community of thousands of families now lay scraped to the foundations of those homes.

As I walked amongst the tile floors and piles of bricks, I thought about the women of the PKK. In a familiar refrain from our time with the teachers in Lake Toba, many of the women said their greatest desire was to learn more about trauma so they could help their friends and family feel better. This outward focus on helping others overcome their trauma sent a powerful message about how this community would recover.

Just as the waves of the tsunami surged through the streets of the city causing devastation and destruction, ripples of healing would now spread across Meulaboh as one person reached out to the next, offering knowledge, guidance, and hope.

As we watched the sun set over the peninsula, the fiery sky illuminated the sea as gentle waves came upon the shore and retreated. My time in Aceh was once again drawing to a close, and soon it would be time to say goodbye. As Nicola and I watched the sunset surrounded by new friends, I thought about the people we had met. Their stories stirred my soul. Their resiliency left us amazed. Their friendship would never be forgotten.

I thought about the fishermen, the mothers, the children, and the countless others who called this city home. Their journey taking them from being victims of a disaster to survivors looking to the future. Inspired once again, I felt gratitude for being called to bear witness to this remarkable transformation.

LESSONS FROM ACEH
focus forward

By focusing on helping others, we help ourselves...

One of the most inspiring aspects of both trips to Aceh was the almost universal emphasis by survivors on helping others. Whether it was their own family, their friends, or their students, so many survivors focused on the needs of those around them. This created an atmosphere of healing in the community. Through this focus on others, they gained knowledge and skills that helped them cope with their own trauma.

Take one step at a time...

The scope of devastation was overwhelming. The amount of debris alone was daunting. And yet, the relief workers and the people of Meulaboh did not let that stop them from starting somewhere...anywhere. So they cleared away the debris, one small area at a time.

We all leave a mark...

Every person who gave to tsunami relief left their mark on Aceh. Each dollar donated represented hope, and people around the world are now a part of this communiity forever.

Photo by Roseann Marchese

the beach

Sunday, September 25, 2005

We have such immense *respect* for those who responded to the call to serve full time in Indonesia. Giving up the comfort of their own lives, they live and work amongst the same devastation as the survivors, *walking alongside them* in the long road of *rebuilding* their community and their lives.

Date:

September 25, 2005

Time:

Afternoon

Place:

Meulaboh, Indonesia

Setting:

After a week of trauma counseling and training, Thomas and Nicola enjoy a day of rest at the beach with relief workers from Indonesia and around the world.

 ## packrat

Three days prior to our departure for Aceh, our entire dining room lay covered with stuff spread out over every inch of available space on the table and the floor. Clearly, some things would not make the cut.

"Really Thomas, I think you should focus first on the essentials, then look at what else you want to bring because there is no way all of this is going to fit," Nicola said, gesturing at the piles of items before us.

Surveying the scene, I noted everything from granola bars, to a headlamp, to our training materials, to gifts for the staff and survivors we would meet.

I resisted the urge to use my "it will fit" packing method from the Marine Corps. Instead, I carefully sifted through each pile, reluctantly weaning out things I thought we could live without. As I came to one pile, I picked up a baseball and glove. Slipping on the glove, I started tossing the ball into the molded center by habit, smiling at a memory from several months back.

Also echoing in my ears was a comment from our team leader, John. "In addition to going over to work with the survivors," he said, "we're also going to encourage the local staff in Meulaboh."

Glancing over at Nicola as she carefully rolled clothes and tucked them into our bag, I knew she was right about limiting what we brought. I also knew, no matter what we cut, the baseball and two gloves needed to come along.

relief

Back in February, during my first trip to Banda Aceh, I had also packed my baseball and two gloves, anticipating the relief workers' need for a little R & R.

During a break from our work, several young Indonesian relief workers and I

headed to the beach. When we arrived at the beach, my new friends could not wait to try on a real baseball glove and play the game they had heard about but never experienced. One of the relief workers smelled the leather and pulled the fingers and webbing back, testing the flexibility and resistance, before placing her hand in a glove for the first time. Another ran down the beach to receive the first pitch, hitting the center of the glove with his free fist after he stopped, just like in the movies.

It turned out he ran a little far, underestimating the shoulder strength it takes to hurl the ball. The first throw landed with a thump in the sand, halfway between the two players. Closing the gap immediately, they began lightly tossing the ball back and forth, catching it more often than not. Dedicated from the start, they jumped high for throws destined to sail over their heads, advanced to pluck low throws out of the sand, and dove with the determination of a Major League player when the ball went out of reach.

Then the inevitable from such unbridled enthusiasm happened. As one of the players tried to catch the ball, it bounced off her glove and popped her in the face. The other player, an observer with a towel, and I rushed over at the sight of copious amounts of blood

pouring from her lip onto the sand. As we crowded around, she started crying.

"Are you okay?" I asked, fearing the worst and feeling a little bad since I brought the cause of the injury.

Looking up at us through the blood and the tears, she said, "I don't want to stop playing. I love this game!"

Smiling in relief, we encouraged her to put the game on hold and have her lip looked at. It turned out the hole through her lip needed five stitches!

She was a great sport about her injury. That night, in good baseball fashion, I had the whole relief worker team sign the ball. We presented it to her with great ceremony. She smiled through it all, even with her swollen lip.

local flavor

Several months later, on our last full day in Meulaboh, Nicola and I

Acehnese women smile from the kitchen of the "SP Cafe" in Meulaboh, where they cook up, among other things, a delicious cheesecake!

Photo by Vicki Oei

Source: "Tsunami Report," Samaritan's Purse

Relief Spotlight

Samaritan's Purse
Serving Hope

Almost immediately after the tsunami, Samaritan's Purse began mobilizing the largest disaster response in the history of the organization. Their efforts in Aceh began with attending to the basic needs of the people, including the need for safe drinking water.

The saltwater brought inland by the tsunami destroyed hundreds of wells, and the threat of disease in the early weeks and months was significant. Samaritan's Purse distributed millions of water purification tablets, ensuring survivors access to a clean water supply, and worked to clean and repair wells. The organization also provided supplies for temporary housing, food, and healthcare.

To assist residents in rebuilding their lives, Samaritan's Purse set up a vocational training center in Meulaboh. The center provides counseling services and also offers tailoring and cooking classes. The "SP Cafe" is a restaurant operated by local women who have learned to prepare Western dishes like pizza and cheesecake. The Cafe caters to the community of international relief workers and provides a livelihood for the women.

Relief Spotlight

Salvation Army
Building Hope

The Salvation Army was one of the first organizations to come to the aid of the Aceh region after the disaster. Initially providing disaster relief through medical care and comfort to the survivors, their efforts soon expanded into helping the region recover. Programs have included livelihood, cash-for-work, and construction projects.

On the one-year anniversary of the disaster, representatives from the Salvation Army gathered with government representatives and residents of Meulaboh for the ceremonial dedication of 500 houses built for families. These yellow houses, built on or between the foundations of homes destroyed by the tsunami, were made possible through funds donated to the Salvation Army in the United States, Canada, the United Kingdom, and the Netherlands.

An aerial view of the yellow Salvation Army houses.

Source: www.SalvationArmy.com

got ready for the relief workers' weekly Sunday afternoon trip to the beach. We were glad to have the opportunity to spend time with our new friends just having fun, not worrying about schedules, trainings, or trauma.

Both of us marveled at the dedication of the relief workers we met, both Indonesian Nationals and foreigners. Called to a ravaged landscape and to some of the most intense human needs one could ever find, they served with passion and purpose.

During our few days in Meulaboh, we got a glimpse into how they take care of themselves on a week-to-week basis, making it possible for them to do the challenging work every day.

A popular activity for the relief workers was watching DVDs of movies and television shows on their computers. Some Sundays they would have all day marathons of shows like *24* and *Lost*, providing both an escape and a chance to catch up on American culture.

The night before, we experienced "dining out" in Meulaboh, getting an unexpected

Photo by Rodney Rascona

taste of home. The "SP Cafe," operated by Samaritan's Purse as part of a vocational facility in Meulaboh, offered a menu with familiar Western dishes like burritos, pizza, and cheesecake. The Cafe provided practical experience for the women who worked there and a "Western fix" for the community of foreigners. We enjoyed a time of fellowship with the local staff. We also enjoyed the food, which, considering the limited availability of ingredients, was quite good.

Before our beach trip, we enjoyed a more local flavor at *Mie Kocok*, the favorite noodle restaurant of several of the staff. Along with plates of noodles, several in our group enjoyed the avocado and chocolate shake, an unusual, but tasty combination. With our stomachs full, now it was time for a bit of fun.

fun & games

As we bumped along on the still-damaged roads, I thought about the universality of our activity that day. Just like in Banda Aceh, the beach was a place of rest and relaxation for the relief workers. A place to recharge on Sundays, their one day of rest. Outside of the privacy of their own dwellings, the beach was also the one place they could be away from the constraints of the traditional Muslim society they lived in. There, they could wear shorts, play, and have fun without the worry of unintentionally offending those around them.

We headed for the agreed upon congregating point for the staff and visitors of several relief organizations. As we drove, we could see areas around the beach that used to be residential neighborhoods. A thick layer of tropical grass only partially obscured the remains of gray concrete slabs and crumbled brick walls. In contrast with this grown-over devastation, in some areas, small yellow houses sprouted up here and there. We learned that the Salvation Army had built these homes for survivors.

Arriving at the beach, we offloaded our chairs and blankets to set up a temporary camp. Even though the heat of the day had passed somewhat and the ocean breeze cooled the air slightly, the ice-cold sodas and bottles of water were refreshing.

Several people from the group went straight for the ocean, playing in the waves. Down by the surf, a palm tree trunk was lodged unnaturally in the sand. One end lay buried, while the other end hovered a few feet off the ground. This created a springboard to jump on. Groups of people took turns lining up on the trunk and bouncing with all their effort until everyone went flying off into the sand. As with all things in Aceh, even this activity reminded us of the reason each of us was there. This palm and several others nearby were clearly dislodged by the force of the water.

I set out my baseball and gloves on the towel next to me, along with a Frisbee. One of the national workers wanted to learn to throw the Frisbee, so one of our team members showed her. We teased her just a little as her throwing style included a small backwards kick at the end of each toss. I was happy she was having so much fun.

having a ball

When you set foot outside the United States, you come to understand the deep passion the world has for soccer. Given this worldwide interest, it wasn't surprising when a soccer ball appeared on the beach. The small group of players grew when several locals from Meulaboh joined

Relief workers watch the sun go down during a Sunday afternoon beach excursion on the outskirts of Meulaboh.

Photo by Mark Blomberg

Chandra Manalu

Chandra Manalu, a relief worker from Medan, Indonesia, spent fourteen months serving in Meulaboh with Food for the Hungry. He reflects on the challenges, as well as what he took away from Aceh.

"Living as a foreigner is a challenge in Aceh and living as a Christian foreigner doubles the challenge. As relief workers, living side-by-side with the people we helped made us more aware of our way of living and our behavior. Eventually, the Acehnese started to acknowledge that I am a Christian and I have values. They said, 'Where are you going to celebrate Christmas? Stay in Aceh...you can celebrate it here.'

One thing that I will always remember about Aceh is our local support staff. They were not only co-workers for me, but also family. They gave so much more than I expected. When I was hungry, they brought me food to eat. When I came home tired, they would sit and talk with me.

Our local staff set an example for me, that as isolated people they could accept me, a total foreigner with different beliefs. They taught me to accept others as they are."

the relief workers. Two teams formed for a friendly, cross-cultural, inter-religious pick-up match, if you will.

While likely not on the minds of any of the players, as a casual observer, I could not help but attach just a little significance at the sight. Western relief workers, national relief workers, and local Acehnese people enjoying such an innocuous activity together, no longer divided by race, culture, or faith.

After the players wore themselves out and the match broke up, one of the relief workers from the United States zeroed in on my baseball and gloves and came over to where I was sitting.

"Would you mind if I throw the baseball around?" he asked, his eagerness evident.

"Go right ahead, that's why they're here," I said.

Saying thanks, he tossed a glove to a friend and took off down the beach to play catch, showing just as much enthusiasm as my Indonesian friends in Banda Aceh. Fortunately, this time there were no major accidents requiring stitches, only a few wild throws that bounced off the backs and heads of innocent bystanders.

As I watched him in the distance, I could almost see the many years he played baseball as a kid. How the feel of the glove and the soundness of the ball landing in the web took him back. Back to his childhood, back home to America.

"Thanks a lot," he said, returning the ball and gloves with a big smile on his face. "It's been forever since I've thrown a baseball...that made my day!"

"Glad I could help," I said,

encouraged that I'd brought a small piece of home to such a challenging place to work and live. Nicola and I exchanged knowing glances, glad that making room for the baseball and gloves was the right thing after all.

When the sun started retiring and the day's activities dwindled to their natural conclusion, we gathered our belongings to return to the guesthouse. As we drove, I reflected on the dedication of all the relief workers and their willingness to give up the comfortable and familiar to serve the survivors of the tsunami.

I found myself trying to put my own day-to-day life in perspective, hoping their example could serve as a reminder not to take my own surroundings for granted. On this day, just as on the day months before in Banda Aceh, I felt privileged not just to serve alongside these remarkable people for a short time, but to play among them as well.

LESSONS FROM ACEH
create community

Find fellowship...

Wherever you are, build a support system. Having people in our lives to connect with counteracts burnout and makes us more effective at tackling the difficulties of life.

Rest and relax...

Even with the chaos of disaster and the influx of relief workers from all over the globe, Aceh remained a strict Muslim culture. For the relief workers, both from other parts of Indonesia and from other countries, this meant living and working in a community with very different customs from their homes. Their weekly trips to the beach became the time and place where they could relax.

Every once in a while, escape...

Like most things, when taken to the extreme, escape is unhealthy. In moderation, however, it can be a healthy coping mechanism. The relief workers' escape into watching TV series and DVDs gave their minds a rest from the constant work and allowed them to escape into a world completely detached from their surroundings.

Into the Heart of Disaster:
Profiles of Service

"Hi Thomas and Nicola, it's John Frick. Could you give me a call back? I have a question for you."

Nicola and I both stared at the answering machine, then looked at each other. Were we being called (literally) to return to Indonesia? A short while later we learned that Food for the Hungry actually wanted to invite Nicola's mother, Vicki, to join a team leaving in just one month. We simultaneously breathed a sigh of relief. Even though we would both jump at the chance to return to Indonesia, the timing was not ideal for us.

We were excited for Vicki's journey back to her country of birth, an opportunity to come full circle and to serve.

The team she joined would focus on education and a livelihood project with the local women's organization. Vicki's Indonesian language skills would prove to be a valuable asset in connecting with the women.

Photos courtesy of Vicki Oei

Coming Full Circle

by Vicki Oei

It was a great surprise when I received the invitation from Food for the Hungry (FH) to join the team from Phoenix to go to Meulaboh. While I grew up in Jakarta, Indonesia, this would be my first trip to Aceh, and the first time seeing the devastation with my own eyes.

arrival

We flew from Medan to Meulaboh on a Mission Aviation Fellowship plane. Driving into the city, we saw new houses being built, but we also saw families still living in tents so many months after the disaster. Despite their living conditions, the adults and children waved, smiled, and greeted us as we passed by.

devastation

Our first glimpse of the utter devastation Meulaboh experienced was a visit to the peninsula at sunset. It really took our breath away to see such devastation on one hand, and on the other hand, to witness the most beautiful sunset. It felt as if God was comforting us with His creation. We went to bed that night blessed to be able to sleep in a modest hotel, even with the noise from a small air-conditioner that cycled on and off throughout the night, lights that flickered, and yellow water to take a bath. *Selamat Tidur! (Sleep well!)*

sunday

On Sunday afternoon we picnicked on the beach, a fun weekly event for the staff in this unrelenting environment. Just as we were ready to go back to the hotel to rest a bit, the Camat (mayor) invited us to a "Cultural Evening." Well, it turned out that the evening was to honor City Councilwoman Peggy Bilsten for what she has done for the people of Meulaboh! Banners everywhere expressed their appreciation.

our work begins

Members of the women's group received baking grants, including ovens and mixers. Our role was to give them an edge in the marketplace by teaching them new recipes for biscotti, banana muffins, and cinnamon rolls/bread. We showed them how they could improvise with local ingredients. Everyone seemed to enjoy the classes and they all wanted us to taste their baked goods! *Selamat Makan! (Bon Appetit!)*

It was very special for me to communicate with the women in Indonesian. It did not take long for them to open up and share stories about their families and their tragedies. I was sad to hear their stories and at the same time glad that they had picked up the pieces and continued to live with hope. Some of them remembered Thomas and Nicola from their trauma counseling workshops for the PKK, which was a nice connection with them.

a special evening

On the last night in Meulaboh, the Camat and Ibu Eva invited us to their home for dinner. It was a special time. I presented them with a windbell made by Paolo Soleri (a renowned Italian artist who lives in Phoenix) in special commemoration of the tsunami tragedy.

the impact

One highlight for me was visiting people who had received grants from FH to rebuild their livelihood. We met Bustami, a thirty-four-year-old man who lost twenty-eight members of his family, leaving only him and his eleven-year-old nephew. He had just opened a kiosk four months before the tsunami. At first, he joined the FH Cash-for-Work program. Then he applied for a grant to reopen his kiosk. Now Bustami has his kiosk, where his nephew helps him after school, and he drives a becak for a second income.

There is no doubt in my mind that each survivor is a miracle. Of course, there are also some problems. Tragedies have a way of bringing out the best and the worst in man. Overall though, what we see are hearts being changed. The people of Aceh ask why the people they thought they hated and despised before would want to help them now? They cannot help but see how many nations and people have come together to offer not just funds but also their labor of love. When we give to relief causes, sometimes it is difficult to know the impact of our gifts. It was a special opportunity to see first-hand how the lives of these survivors had changed because of the help of people who live so far away.

Please continue to pray for the people in Aceh and for those who serve there. *Tuhan memberkati! (God bless!)*

the visit

Wednesday, January 11, 2006

I witnessed something truly **amazing** *this week. Just as I stepped into their world, two people from Aceh have now stepped into mine. Brought* **together** *once again, we celebrated* **friendship** *and understanding between our cultures. And all of us were blessed.*

Date:

January 5, 2006

Time:

Evening

Place:

Phoenix, Arizona

Setting:

Thomas and Nicola are part of a welcoming committee for the leader of the city of Meulaboh and his wife during their first visit to the United States. They came to give thanks on behalf of the people of Aceh.

Special thanks to Jayme West of KTAR FM and Tricia Johnson for use of quotations from their interviews with the Camat.

commemoration

They gathered on the peninsula. Exactly one year had passed since towering waves came crashing over this section of land jutting out into the sea. A year since thousands lost their lives. A year since their city lay in ruins, and they felt sure that it was the end of everything.

It was a year in which their community, closed for so long to the outside world, saw dozens of countries come to their doorstep offering food, water, medicine, and aid. A year in which they opened the door and learned that perceptions are sometimes inaccurate and that hearts can change. A year in which they formed friendships rooted in a collaborative effort to rebuild. A year in which regaining hope became a necessity for recovery.

A year after their lives changed forever, they gathered on the land that was scraped clean by the tsunami. In the midst of lush green growth that softened the rough edges of destruction, they set up easels, canvases, paints, and brushes.

Looking out across the land where so much was lost, to the sea that for a year had stayed within its normal bounds, they began to paint, commemorating all that was lost and celebrating all that was gained.

As the paintings came together on canvas, more than one had a destiny to travel far across the sea. Teuku Dadek, a leader in his community, would soon embark on a journey with his wife to a land far different from his home. He would bring words of thanks on behalf of his people and he would carry the work of the painters as gifts for new friends.

As the painters worked, and he looked on, perhaps they wondered... now that a year has passed, *will we be forgotten*?

arrival

We stood in a row at an America West Airlines gate in Terminal 4 of Sky Harbor Airport in Phoenix, Arizona. Each of us craned our necks to try and catch a glimpse of our arriving visitors. Dozens of travelers filed past us off the plane, heading straight for baggage claim, probably quite unaware of the great occasion of their flight's arrival. If they looked around as they walked by, they might have recognized the mayor of Phoenix, waiting along with the rest of the welcoming committee. They might have seen the reporter with her media credentials and the photographer, ready to capture the moment for posterity.

As we waited, many of us knew what the travelers had just been through, having made the same long journey before. Only for them, after over thirty hours of travel, five flights, ten time zones, and pure exhaustion, they would be greeted by a huge welcoming committee, including a bevy of video cameras and reporters just outside the security checkpoint. I could not imagine what that would be like!

We were still amazed that they were coming at all, as there were many barriers that could have prevented their visit. Just a year before, it would have been virtually unthinkable for Indonesian citizens from Aceh to get visas to visit the United States. For this occasion, it was the recommendation of Senator Jon Kyl and Mayor Phil Gordon that made it possible for the U.S. Embassy in Indonesia to issue the visas. On top of that, new passports were needed since their old ones were swept away by the tsunami.

Their visit also meant taking time from their responsibilities of leadership in Meulaboh and their newly formed family. This was the first time they would leave their children, only six months after adopting

their niece and nephews who became orphans of the tsunami.

Despite all these obstacles, it was important for them to make the journey. After months of planning and paperwork, Teuku Dadek, the Camat (cha-maht) or mayor of Meulaboh, and Ibu Eva (ee-boo eh-fa), his wife, were coming to the United States.

They were coming to see first-hand a city that reached across the ocean to care for their people. Coming to experience a culture as foreign to them as theirs was to us. Coming to bring tidings of thanks from people now full of hope.

As we watched the weary, yet smiling visitors step off the plane, I wondered what this week had in store for them. Would they be as changed by stepping into our world as I was stepping into theirs? What would experiencing American culture be like for them? And what would they tell their people when they returned?

gathering

The following morning, a bright and sunny January day, we drove to a revitalized warehouse building just south of downtown Phoenix. This was an example of renewal in our own city.

The arrival of the Camat and Ibu Eva from Indonesia was an exciting event, bringing out officials, friends, and the media to greet them. After an on-the-spot interview with the local media, they were ushered off to get some rest after their long journey, as more media appearances awaited early the next morning.

City of Phoenix Mayor Phil Gordon, the Camat, and Food for the Hungry President Ben Homan walk through Terminal 4 in Sky Harbor Airport.

The Camat is interviewed at the airport.

Photos by Bob Rink

GATHERING

The first anniversary luncheon celebrating the partnership provided a special opportunity to recap the progress made in the first year of the partnership.

Pat McMahon, Ben Homan, the Camat, Peggy Bilsten, and Ibu Eva gather around the sculpture of a Phoenix rising from the ashes.

While we had worked with Ibu Eva in Meulaboh on the training for the PKK women's organization, their visit was actually the first time we met the Camat in person. Like so many others who have a connection to Meulaboh, we are now honored to call them friends.

Photos by Bob Rink

The Bentley Projects, a combination art gallery, cafe, and event space that was reclaimed from a rundown state, would serve as the site of a commemorative luncheon to celebrate the one-year anniversary of the adoption of Meulaboh by the City of Phoenix. The event was a fundraiser, a forum for the Camat to say thank you on behalf of his people, and for us, a reunion with our team from the September trip.

Before the luncheon, Nicola's parents and two close family friends, all immigrants from Indonesia, talked with the Camat and Ibu Eva. They seemed to enjoy the opportunity to converse in their native language, especially Ibu Eva, who spoke very little English. Nicola and I set up a few mementos for the guests at our table to look at. They included a copy of the issue of *East West Magazine* with a report from my February trip, a copy of our *Restoring Hope* curriculum, and a photo album with pictures from our September trip.

A video by Food for the Hungry recapped a year of work in Meulaboh. Included in the video was a short clip of an interview with Nicola and me at Lake Toba. In the clip, Nicola commented on our appreciation for the *Rising to Help* partnership's ten-year commitment to walking alongside the people of Meulaboh, even after the story faded from the limelight. The video also showed school children in Meulaboh hokey-pokeying with smiles and laughter, as our "instruction" with the teachers at the retreat trickled down to the classroom.

After lunch was served, the Camat came up to speak. He took the opportunity to personally thank the many people and corporations who supported the relief effort. As he shared the accomplishments and the challenges of the last year of rebuilding, it struck me just how much responsibility lay on his shoulders. He and his wife had set aside their own grief to work for the good of their people. Finally, overcome with emotion, he stepped down from the podium.

A presentation to the Camat of a small sculpture of the City of Phoenix firebird rising from the ashes completed this coming together of representatives from two governments, two cities, two countries.

whirlwind

For the Camat and Ibu Eva, their visit was not only a chance to say thank you, it was also a crash course in American culture. At the luncheon, they were presented with Phoenix Suns basketball jerseys, and an invitation to attend a game during their visit.

At the game they wore their new jerseys and enjoyed the fun atmosphere. They even got in on the action when the Phoenix Suns Gorilla tossed tee-shirts into the crowd (one actually dropped right on Ibu Eva's head…she got to keep it!). Asked about the experience of attending the game, the Camat said they really enjoyed it, commenting that player Steve Nash is "one of the stars for me…a very good player!"

During their visit, they took a side trip to Kentucky to visit two churches that also supported the relief effort. While in Kentucky, the Camat spoke before 8,000 people in four church services. He shared his first-hand perspective of the effect of their donations and thanked them for their support. For their visit, the church invited people from the local Indonesian Muslim community for a reception, another cross-cultural and inter-faith bridge from the partnership.

The Camat enjoyed the scenery of Kentucky, commenting that he liked seeing the ranches. During a visit to a country-western store, he met "real" cowboys and even bought a cowboy hat as a souvenir.

Food was an adventure for them. The Camat described how at home they eat the same thing every day. Their meals consist of rice, oil, chili, and fish, three times a day. In the United States, they experienced culinary diversity. "Here you can eat many different things, and the portions are very big," he said with a smile, commenting that the appetizer portion was big enough for him.

Asked how his opinion of America had changed, the Camat shared that before, most of his information about American culture came from television and Muslim magazines and newspapers. From his visit, he understood that his perception of Americans was skewed. During an interview, he shared, "It's very interesting for me…so many things I learned. When I go back to my people, I will tell to my people, 'The United States is not like you're thinking. They are similar to us. They love the church, they love their religion, same way with us, too. Everywhere, people around the world are similar.'"

I was particularly impressed with

the Camat's comfort in front of the media, especially since English is not a primary language for him. From the moment he stepped off the plane and throughout their visit, he spoke to audiences and television, radio, and print media, sharing his thanks and his impressions of America. He attended a Phoenix City Council meeting, business luncheons, and was a featured guest on a FOX TV morning news program. Especially meaningful

Photo by Bob Rink

At the luncheon, the Camat and Ibu Eva received gifts of Suns basketball jerseys and an invitation to attend a game.

They enjoyed the game, even catching one of the Phoenix Suns Gorilla's coveted tee-shirts.

Photo courtesy of Pak Dadek

was his visit to school children who raised money for Meulaboh. "The children were taught to care," he said. Everywhere he went, the Camat shared the grateful heart of his people.

a gift

In most societies it is traditional to gather for meals to celebrate special occasions. At the end of the Camat and Ibu Eva's visit to Phoenix, we gathered for an Indonesian feast. After a whirlwind week of new people, lots of English, and new food, our guests had the chance to eat familiar cuisine and to speak their language with Nicola's parents and their Indonesian friends. We all enjoyed sharing Indonesian food with the American guests, warning them to be cautious with the *sambal* chili sauce!

That evening, the Camat helped put the significance of his visit in context. He described how during the first Gulf War, most of the Acehnese sided with the Iraqis, hoping they would be victorious. Now, however, many of his people have a much better understanding of the heart of the American people. "You did for us what we would have never done for you before," he told me. Over and over, throughout their trip, he commented

After their whirlwind visit, it was nice to enjoy one last evening together before our visitors returned home. Vicki, Ibu Eva, and Nicola pose with the Phoenix photo album.

that it was amazing to them just how far Phoenix is from Meulaboh. They marveled at how much people cared from such a great distance.

Toward the end of the evening, we exchanged gifts. Nicola and I gave the Camat and Ibu Eva a Phoenix-themed photo album, with pictures from their visit, as well as picture books about the desert, so they could show their children what they experienced in Arizona.

Then, the Camat said he would like to present us with a gift, something from their home. Picking up a large cardboard mailing tube, he recounted how their community commemorated the anniversary of the disaster. On December 26, 2005, a group of artists gathered on the peninsula at the edge of the city, the place where so many thousands of people lost their lives. The artists set up their easels and, side-by-side, painted for those who were lost.

Unrolling a rectangular canvas,

the Camat revealed a work painted by a friend of his. In bright colors, it captured the spirit of the event. The scene shows the ruins, slowly being taken over by new growth. It also shows the other artists at their canvases, painting their own interpretation of what they saw. Most touching, the painting shows small children huddled between the towering palm trees, watching as the painters work.

Overwhelmed, we saw the scene through the eyes of the artist, the tragedy of the past alongside the possibility for the future. The hope for what lay ahead for his city and its people.

A year and a half later, the painting hangs in our office, a daily inspiration as we finish this book, which is our own way of commemorating tragedy and celebrating hope. We decided that the image is a fitting way to close out this story of transformation. With it, we hope to send a message to the people of Aceh… *you are not forgotten.*

LESSONS FROM ACEH
give thanks

Express gratitude…
Expressing gratitude is important. It was a long journey and a logistical challenge for the Camat and Ibu Eva to travel to the United States, but they wanted to thank their supporters personally. It was meaningful for those who supported the relief effort to hear a message of thanks directly from a representative of the community they helped.

See for yourself…
If you have the opportunity to experience something, anything, first-hand, take it. The perspective and connection that each of us involved in the tsunami relief effort gained by traveling to and from Meulaboh are immeasurable. It is truly a life-changing and eye-opening experience to step into a completely different world.

Keep an open mind…
The Camat's comments about the change in his perception of Americans reminded me that my perception of the people of Aceh probably changed just as much as a result of our interactions. There is always more to be learned. A personal connection is key to greater understanding.

We must not forget…
Disasters are fleeting in the public eye. They captivate while in the news, but fade as soon as the next story breaks. As work continues in these places, it is important to recognize that the road to recovery is long and that support is still needed after the spotlight fades.

EPILOGUE

As with transformation, the story continues…

It has been two and a half years since a 9.3 earthquake created the largest natural disaster in recorded history.

Thirty months since the tsunami destroyed families, livelihoods, and communities.

Just under one thousand days since the world came together in Aceh to help heal and rebuild.

Hope returned and life goes on in Aceh.

ACEH

from civil war to peace – Prior to the tsunami, Aceh was in the midst of a civil war that had lasted thirty years, and all previous attempts to find resolution between the Acehnese separatists and the Indonesian government had failed. After the disaster, in August 2005, the two sides came together and signed a peace treaty. While there is still much to be done to continue the peace, it was a hopeful development for Aceh.

In August 2007, a team from Phoenix conducted a conference in Meulaboh on good government practices. Over 500 law enforcement and government officials attended workshops designed to help strengthen the community, an impressive turnout given the region's very recent political instability.

from disaster to rebuilding – Many organizations remain in Aceh, walking alongside the communities they have helped since the disaster. Their five- and ten-year plans for supporting the rebuilding effort are encouraging, especially as they occur out of the spotlight.

In the community of Calang (cha-lahng) north of Meulaboh, Food for the Hungry has been working with farmers on agriculture projects. This includes showing the Acehnese new methods of organic farming that use fewer resources, but yield bigger crops. The farmers were skeptical at first, so two women volunteered to try the new method. With their success with the organic methods, all of the farmers quickly converted! This program, which solidifies the survivors' ability to earn a living and results in a healthier community, is only one example of the many programs in operation today in Aceh.

from trauma to hope – At the end of the teachers retreat, Bethany and the teachers presented us with gifts of appreciation. Nicola received an embroidered purse, while a teacher presented Thomas with a traditional Acehnese knife. It wasn't until later, though, that we received the following "souvenir" from our trip, a quote from a teacher who attended the retreat:

"We feel like we can't go on living at times because our whole lives were destroyed…our feelings were dead, our thoughts were dead, we had nothing to feel anymore. But you have come to help us. And we are ready to have joy again. We are ready to have joy in our classrooms and in our hearts."

We are hopeful for the future of Aceh.

TELLING THE STORY

the story — Our time in Aceh led to diverse opportunities to share about our unique experiences working with survivors and relief workers. In 2005 and 2006, *East West Magazine* featured journal excerpts from our trips.

In the months after Thomas' return, we presented *Voices of Aceh: Echoes of the Tsunami*, a dramatic multimedia presentation of our experiences and the accounts of the survivors, to churches who provided support for our travels.

In 2006, we had the opportunity to speak about our time in Aceh at two international conferences. In Brisbane, Australia, we presented our trauma counseling work with survivors, while in San Diego, California, we spoke on the transformation of Aceh from closed society to collaborative, international community.

the book — This book represents the culmination of all those experiences; the positive response from readers and audiences encouraged us to pursue telling this story in a more durable format. We chose to publish independently to retain editorial oversight and to ensure that our vision for a visually engaging storytelling format could be maintained. While this has meant that all responsibility and cost for the project has fallen on us, it has allowed us to keep the focus of this project on telling the story and raising funds for the relief effort.

The stories in *Transformation from Tragedy* are not meant to represent everything that happened in Aceh. We realize that the relief and rebuilding effort is not perfect. By and large, though, our experiences with relief workers and survivors were positive and inspiring, and the stories reflect this. Our story is only one of thousands, but we hope it provided you with a glimpse into a much larger world of caring and healing.

the future — People often ask if and when we will return to Aceh. Our response is that both times in the past, we've only had about a month's "notice" for our departure. We choose to trust that when the time is right, we'll know and will be made ready. We hope that our call will one day lead us back to Aceh, but we are patient, knowing that for the last year our purpose has been to finish this book.

In the end, we are grateful that we were called to this unique and life-changing experience. It has been our privilege to share this story with you.

Thomas & Nicola

P.S. *We would love to hear from you. Visit us at www.WordPointPublishing.com or email us at info@wordpointpublishing.com.*

Dedication

We dedicate this book to the survivors and relief workers of Aceh, Indonesia.

To the survivors, we thank you. You trusted us, though we were strangers. You shared your stories, though we could not possibly fathom what you had been through. You helped us truly understand the depths of resilience in the human spirit, which we could not have imagined. It is our sincerest hope that our telling of your story is worthy of your walk.

To the relief workers who serve, your heart for the people of Aceh inspired us to see this project through. We hope that through your stories, the world will better understand the potential for compassion to change lives and impact the world.

& Acknowledgments

We would like to thank all those who contributed, tangibly and intangibly, to the creation of this book and the telling of this story.

Our earliest supporters were our **family and friends** who, when this whole idea of going to Indonesia came up, did not tell us we were crazy but supported us unfailingly. Special thanks to Nicola's parents, **Okky & Vicki Oei**, whose contacts created the first link between us and Aceh, and to Nicola's grandfather, **Opa**, who is our number one fan and bought the very first book.

For **everyone who supported our two mission trips**, we are incredibly grateful. You prayed tirelessly for the journey and the work. You gave financially for everything from training materials to plane tickets. (You even took us shopping to REI for the ever-present yellow, Hawaiian shirt—thanks Sandy & David!) Without all of you, our travels, and therefore this book, could not have happened.

Thanks go to each of the people we encountered on our two trips to Aceh, many of whom you met within the pages of this book. From Thomas' February trip, special mention goes to **Tor & Fera Torsina**, who cared for Thomas like he was family, and **Peter**, who extended the first invitation and became a true friend. We would also like to acknowledge the Catholic and Methodist churches in Medan and Banda Aceh for coordinating trainings for relief workers.

From our September trip, our appreciation goes to **Heidi Blomberg, John Frick, Bethany Nanulaitta,** and all of the Food for the Hungry staff who worked so hard on planning and executing the retreat and trainings, making our work there possible. We want to acknowledge the members of our team, **Gordon James**, **Meredith Lewis**, **Roseann Marchese**, **Pat McMahon**, and **Rodney Rascona**, who not only became great friends, but who were so instrumental in the success of the retreat. We also thank **Peggy Bilsten**, whose heart for the people of Meulaboh inspires, and **Teuku Ahmad Dadek & Eva Mahnizar**, who represent the heart of their people so very well.

Throughout both trips, we relied extensively on **translators** to help us convey information and to hear the survivors' stories. Our thanks to the many people who were our voices and bridged the barrier of language.

There are many people to thank for the existence of this book. Our appreciation goes to **Anita Malik**, publisher of *East West Magazine*, who originally ran the magazine stories that inspired this book. In August 2006, a "chance" meeting with **Elaine Wright Colvin** transformed a germ of an idea for a book into a fully-fledged concept. We thank you for your guidance, encouragement, and introduction to the "**Glorieta Gang**," a wonderful group of writers. Our most heartfelt thanks to Elaine and the entire gang for welcoming us into your fold and providing more encouragement than we could have ever expected.

We give our thanks to all the **contributors**, identified on pages 153-156, who graciously allowed the use of their words and images to help tell this story. From the beginning, we wanted this book to reflect the diversity and breadth of the relief effort, and we believe your perspectives enrich this story.

We thank especially **Rodney Rascona**, a photographer with a remarkable eye for humanity, not only for the use of your incredible photography to tell this story, but also for your support. On the shores of Lake Toba you encouraged us to think about what unique impact we could make on the world. This book is a direct result of the gentle challenge you put before us.

Big thanks to our friend **Roseann Marchese**. Aceh may have brought us together, but tiny fish, *The Office,* and gelato cemented our bond. We are so grateful for your friendship and your words of wisdom.

Gratitude forever goes to our team of reviewers: **Barbara Albert, Elaine Wright Colvin, Twink & Denny DeWitt, Deanne Kamp, Kelly Hislop, Mary Ann Hislop, Roseann Marchese, Vicki Oei, Kurt & Kristy Ostrem, Sandy Stoner**, and **Helene Lie Tate**. We so appreciate your time and your commitment to helping us tell this story clearly and effectively.

Our thanks to **Jim & Lisa Sipe** of Star Dot Star, LLC (www.StarDotStarDesign.com), who graciously allowed Nicola to practically live at their house during the layout process, and who quite literally transformed our vision for this project from concept into reality one page at a time. We have so appreciated your friendship, expertise, and efforts throughout this very long process.

Thank you to **Gordon James** (the PR Guru), for believing in us and this project from the very beginning.

Our most sincere appreciation goes to former **President George H.W. Bush** and his chief of staff, **Jean Becker**. Your presence in this book is more than first-time authors should even dream about. We are honored to have your foreword as an introduction to this story.

To **our readers**, we thank you for taking the time to learn about the survivors and relief workers in Aceh. While they live and work in a land distant from ours, we hope this book served as a window into their world.

And of course, a woof to **Becky & Spencer**, because no matter where we go or how long we've been gone, you're always happy to see us and welcome us home.

Writing, editing, and publishing this book has been a true labor of love and a significant challenge over the past year. We give our thanks to every person who asked about our progress and encouraged us with their belief that we could do this.

Music was also key in the challenging moments. Specifically, the song "Voice of Truth" by the Casting Crowns brought back into focus time and again the true purpose of this endeavor and the real source of our words. And the lyrics of "Amazing Grace" by Chris Tomlin reminded us that for all the tragedy created by the tsunami waves, mercy still reigns.

Finally, we know without a doubt that this book was made possible only because of **The One**, the Great Comforter, whose love and grace provide the means for us to love and comfort others. It is in His name that we give thanks.

Behind *the* Scenes

The people and relief organizations in *Transformation from Tragedy*

The tsunami relief effort was one of the largest humanitarian efforts in history, with thousands of people and organizations descending on Southeast Asia from throughout the world. They brought supplies, expertise, funds, and, perhaps most importantly, hope.

It was important to us to incorporate the perspectives and experiences of many people who served in Aceh into the stories to provide the reader a broader look at the massive relief effort. In the following pages, get to know the people and organizations you met throughout the book.

Thomas & Nicola Winkel are coauthors and publishers of *Transformation from Tragedy*. Married for eleven years, they are partners in life, business, travel, and writing.

Thomas' Background: Thomas served in the United States Marine Corps for four years, including tours in Japan, the Philippines, and the first Gulf War. He was honorably discharged in 1992. Thomas holds a Master's in Counseling Psychology from the Lewis & Clark Graduate School of Professional Studies in Portland, Oregon. He is a Licensed Professional Counselor in the State of Arizona.

Nicola's Background: Nicola holds a Bachelor's in Psychology from Lewis & Clark College in Portland, Oregon. Her professional background includes social work and program development in the areas of child welfare and domestic violence, as well as writing and editing in the publishing field. Nicola's entire family emigrated from Indonesia to North America in the 1960s.

Current Work: Together Thomas and Nicola own and operate The Waypoint Group, LLC (www.TheWaypointGroup.com), which provides counseling, training, and consulting to individuals, families, and organizations. They also own and operate WordPoint Publishing Group, LLC (www.WordPointPublishing.com), which published *Transformation from Tragedy*. In 2005 they served together in Aceh, Indonesia, and helped the survivors of the earthquake and tsunami by creating *Restoring Hope*, a curriculum and training focused on helping survivors heal and move forward.

The Winkel Approach: Thomas and Nicola firmly believe in living a life interjected with laughter, full of meaningful experiences, and shared in a commitment to something much bigger than the two of them. They try every day to focus their life in this direction.

The organizations featured in *Transformation from Tragedy*:

American Red Cross - *the life*	www.RedCross.org
Food for the Hungry - *a partnership of hope*	www.FH.org
HumaniNet™ - *the call*	www.HumaniNet.org
Mission Aviation Fellowship - *the retreat*	www.MAF.org
Salvation Army - *the beach*	www.SalvationArmy.org
Samaritan's Purse® - *the beach*	www.SamaritansPurse.org
UNICEF - *the postcard*	www.UNICEF.org
USAID - *my presidents*	www.USAID.gov

Rodney Rascona
photography

As a photographer, Rodney works in the areas of conceptual advertising, landscape, and automotive photography. He generously donates his time and skills to the relief community by creating poignant portraits of the human condition around the world. He was in the field immediately after the tsunami, documenting the plight of the survivors and the relief effort (www.Rascona.com).

Photo courtesy of Rodney Rascona

Ira Lippke
into the heart of disaster – a journey of hope

Ira started practicing the art of photography at age fourteen. Since then he has worked in the music industry, shot fashion campaigns and editorial portraiture for designers and magazines, and captured weddings around the world, all with his distinctive vision. He has also traveled the world documenting humanitarian work in Asia, the Middle East, Central America, and Africa (www.IraLippke.com).

Photo courtesy of Ira Lippke

Scott McAlvany
into the heart of disaster – a journey of hope

Scott first traveled to Indonesia while part of a Youth With a Mission program in Australia. It was then that he got involved with an orphanage in Kuta, Indonesia, where he subsequently spent every winter and summer break. A primary focus of his work with the children is preparing them for the future by teaching them accounting and entrepreneurial skills.

Photo courtesy of Ira Lippke

Shane Essert
the postcard

Shane is fourteen-years-old and is a freshman at Brophy College Preparatory in Phoenix, Arizona. He is a varsity swimmer and hopes to play baseball. Shane also swims year-round for the Arizona Marlins U.S.A. Swim Team. He enjoys ping-pong and hanging out with friends. One day he hopes to attend a United States military academy for college.

Photo courtesy of Food for the Hungry

M. "Tor" Torsina
the yes

After a corporate career in marketing in Jakarta, Indonesia, Tor now focuses his efforts on full-time volunteer work with Christian organizations in Indonesia and throughout Asia. He and his wife, Fera, have a son and a daughter, both married, and one grandson, who all live in Arizona.

Photo courtesy of M. Torsina

Peggy Bilsten
into the heart of disaster – a partnership of hope

Peggy Bilsten served on the Phoenix City Council from 1994 to 2007, including two terms as vice-mayor. Peggy was instrumental in establishing the partnership between the City of Phoenix and Food for the Hungry, traveling to Meulaboh just weeks after the tsunami. She continues to be a strong advocate for the people of Meulaboh. Peggy and her husband, Tom, have two adult children.

Photo by Bob Rink

Benjamin K. Homan
into the heart of disaster – a partnership of hope

Benjamin is president of Food for the Hungry, an international relief and development organization serving in developing countries around the world. In his leadership, he has led assessment teams to Baghdad, Afghanistan, tsunami-ravaged Indonesia, Sudan's Darfur region, and other devastated zones of the world. When not traveling, he enjoys camping with and spending time with his wife and three children.

Photo by Bob Rink

John Frick
the team

John is the Senior Director of Ministry Partners at Food for the Hungry. He is passionate about the cause of the poor both in spirit and in body, as well as giving a voice to those who cannot speak for themselves. He has led five teams to Meulaboh for tsunami relief. John enjoys sports and the theatre. He and his wife, Debbie, reside in Maricopa, Arizona.

Photo by Rodney Rascona

Pat McMahon
into the heart of disaster – aceh! (god bless you!)

Pat was born into show business, spending the first thirteen years of his life on the road with his parents' vaudeville act. After college, he settled in Phoenix, Arizona, where he has been on the radio and television ever since. An award-winning talk show host, his honors include seven Emmy awards and four Halls of Fame. He is currently a radio host on KTAR FM and is seen daily on the Pat McMahon Show on AZTV .

Photo by Rodney Rascona

Bethany Nanulaitta
the retreat

While her family is originally from Georgia, Bethany spent her first seventeen years in Indonesia. After the tsunami, Bethany wanted to help people in her other home. She moved to Meulaboh where she was dubbed "Fake Bule," as her outward appearance identifies her as a foreigner or "bule," while her fluency in Indonesian gives away her native past.

Photo by Rodney Rascona

Heidi & Mark Blomberg
into the heart of disaster – a year in aceh

Heidi and Mark met while serving in Meulaboh, Indonesia. Married since 2006, they currently make their home in Phoenix, Arizona. Mark is a pilot and aviation mechanic, and Heidi is a relief coordinator. They plan to join Mission Aviation Fellowship and serve together in the field.

Photo courtesy of the Blombergs

Gordon James
the city

Gordon is the founder of Gordon C. James Public Relations, a full-service media relations, event management, and governmental affairs agency. Prior to forming the agency, he served on President George H.W. Bush's staff at the White House and was named Director of Special Events for the 54th and 55th Presidential Inaugural Committees (www.GCJPR.com).

Photo by Rodney Rascona

Teuku Ahmad Dadek & Eva Mahnizar
the city, the visit

Pak Dadek is a graduate from UGM law school in Yogyakarta, Indonesia. At the time of the tsunami, he was thirty-seven-years-old and the Camat of Meulaboh. His wife, Ir. Eva Mahnizar, is an agriculture engineer graduate from Unsyiah Banda Aceh. They have three children, thirteen-year-old Popon, nine-year-old Dani, and four-year-old Nisa.

Photo by Bob Rink

Pete Howard
the city

Pete has been serving with Food for the Hungry since 2003. His work has taken him around the world to people suffering from HIV/AIDS and hunger in Africa, the tsunami and earthquakes in Asia, and war and conflict in Iraq and Sudan. Pete is currently a Catherine Reynolds Fellow at Harvard's Kennedy School of Government pursuing a master's with an emphasis in international development and leadership.

Photo courtesy of Heidi Blomberg

Evie Nirwana
the city

Originally from Medan, Indonesia, Evie moved to Meulaboh a few months after the tsunami. There, she worked as a livelihood coordinator for Food for the Hungry, focusing her efforts on helping people start or re-establish small businesses. Her desire is to help needy and hungry people around the world have a better life.

Photo courtesy of Evie Nirwana

Chandra Manalu
the beach

Chandra is originally from Medan, Indonesia. He moved to Meulaboh in July 2005 and spent fourteen months working as a livelihood officer for Food for the Hungry. He is currently pursuing a master's degree at the University of Melbourne in Australia under a scholarship from AusAid. He plans a career in development work.

Photo by Rodney Rascona

Vicki Oei
into the heart of disaster – coming full circle

Born in Jakarta, Indonesia, Vicki immigrated to Toronto, Canada, in 1966. She and her husband of thirty-nine years, Okky, a physician, own and operate a pain relief clinic in Scottsdale, Arizona. Vicki has two grown children (Nicola & Jason) and two grand-dogs (Becky & Spencer). She enjoys gardening and ministering to friends and family.

Photo by Bob Rink

Endsheet Photo Credits

Opening endsheet (left to right, top to bottom): Thomas Winkel, Thomas Winkel, Thomas Winkel, U.S. Navy, U.S. Navy, Rodney Rascona, Rodney Rascona, Rodney Rascona, Vicki Oei, Rodney Rascona, Rodney Rascona, Thomas Winkel, Rodney Rascona, Rodney Rascona, Vicki Oei, Rodney Rascona, Rodney Rascona

Closing endsheet (left to right, top to bottom): Thomas Winkel, Heidi Blomberg, Vicki Oei, Heidi Blomberg, Thomas Winkel, Vicki Oei, Rodney Rascona, Nicola Winkel, Heidi Blomberg, Vicki Oei, Nicola Winkel, U.S. Navy, Nicola Winkel, Rodney Rascona, U.S. Navy, Heidi Blomberg, Vicki Oei, Nicola Winkel

A Closer Look

The Imagery of
Transformation from Tragedy

Early on it was apparent that the story of the tsunami and the ensuing relief effort would best be told with both words and images. While many of the photos come from our collection, we greatly appreciate the many other people involved in the relief effort who allowed us to use their photos to illustrate the story. Here is a closer look at the artwork from the book.

Our thanks to the following people and organizations who graciously provided permission to use their photographs to tell the story of the tsunami in Aceh:

Mark and Heidi Blomberg
Gregg Edgar
GeoEye/CRISP-Singapore
Ira Lippke
Samuel Lippke
Roseann Marchese
Vicki Oei
Rodney Rascona
Bob Rink
United States Navy
White House Photo Office

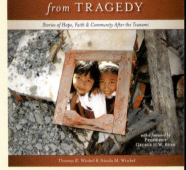

the cover

The primary themes for the cover are transformation and hope. The background image, by Rodney Rascona, is rubble from a peninsula in Meulaboh. The frame provides a "window" into the renewed spirit of hope, beautifully represented by his photo of the children.

voices from aceh – satellite photos

For many people around the world, the first time they truly understood the extent of the disaster was through satellite photos showing the drastic before and after images. These satellite images are courtesy of GeoEye.

voices from aceh – death toll spreads

Rodney Rascona's ability to capture both the scale and humanity of the disaster makes his photographs a powerful account of the tsunami and aftermath. His images, alongside the mounting death toll, draw us from a distant to a grounded view. See more of his tsunami images at www.Rascona.com.

profiles of service

From Ira and Scott's journey, to the compassion of our military forces, to Vicki's full circle experience, these images give personal glimpses into the hearts, minds, and viewpoints of those who served when Aceh needed them most.

story artwork

The oil painting that illustrates the last story in the book, *the visit*, provided the inspiration to create paintings for the opening spreads of the other stories. A special Adobe Photoshop filter transformed photographs into oil paintings. On the following pages, each of the original photos is presented, along with more details on the images.

the call

This photo, from Thomas' first trip to Aceh, illustrates the random nature of the devastation. As the image for the opening chapter, it also sets the point-of-view for the book; this is a story we see through Thomas' eyes.

the postcard

Thomas met hundreds of children during his visit to Banda Aceh and distributed hundreds of postcards. Despite the language barrier, a connection was made, and these little ones enjoyed their gifts from across the sea.

the yes

This image of disaster perfectly frames an Indonesian flag in the distance. Many such flags went up around Aceh, a visual indicator of the resilience of the people.

the boy

The adults and children living in refugee camps were usually excited to meet visitors, especially foreigners. Cameras tend to be people magnets in Indonesia, with plenty of smiles despite the circumstances.

the durian

This story of camaraderie in the midst of disaster offers a bit of levity midway through the book, and a bit of education on the durian itself! Thomas is glad to get the word out on one of his favorite foods.

the one

Finding an image to illustrate this story was a challenge. We felt that this tree, standing alone, was a fitting representation of the theme of the story. One house or tree standing alone amidst rubble was a relatively common sight in Aceh.

the life

The juxtaposition of life and death in this story was a reality in Aceh. This wall of hope was difficult to look at, but it provided an important human element to such an incomprehensibly large death toll.

the team

Thomas looks on as Rodney Rascona films Pat McMahon on the bow of the boat on the way to the retreat at Lake Toba. The footage and interviews later aired on Pat's Phoenix television show.

the retreat

The Mission Aviation Fellowship seaplane takes center stage, as it brings the teachers to the retreat at Lake Toba. Traveling by seaplane provided the opportunity to see a bird's-eye view of the beautiful, mountainous terrain surrounding the lake.

the city

In this photograph of Meulaboh, taken from a moving vehicle, the ragged top of the building provides a reminder of the disaster as people below go about their daily life nine months later.

the beach

This beach outside Banda Aceh served as the baseball training ground from the story. As much fun as Thomas had teaching the game to his new friends, reminders of disaster, like the small shoe on page 57, were never far away.

the visit

This painting is a treasured gift from Pak Dadek & Ibu Eva. The commemorative event in Meulaboh where it was painted was part of an effort to encourage arts in the community.

AND LIKE A FLOOD, HIS MERCY REIGNS
UNENDING LOVE, AMAZING GRACE.

— Amazing Grace, Chris Tomlin